ISCS SandFest 2018

ISCS

September 20, 21, 22, 23, 24
Jacksonville, FL

Edited by Lori Carter & Peter J. Blau

A completely volunteer, non-profit organization whose purpose is to promote the hobby and collection of sand gathered for personal gratification and amateur and/or professional study

All proceeds above cost will be used to further education in Earth Sciences and to support educational efforts of the International Sand Collectors Society.

This edition published by:

The International Sand Collectors Society
PO Box 1786
Lilburn, GA 30048
www.sandcollectors.org

ISBN: 978-1-937617-12-7

Cover:
Menorca, Spain by Leo Kenney
Sand Shadow Box by Fran Peters
Front Page:
Murcia, Spain, by Leo Kenney

Table of Contents

Symposium Chair's Message

It's been a great pleasure to work with so many fine folks during the planning and presentation phases of the *SandFest 2018* symposium! Thom Hopen and Lori Carter have worked especially hard to recommend speakers and work in the background to enable the success of the event. Thom's broad background in ISCS and Lori's Georgia connections have proven valuable in assembling a set of presentations that span a wide range of topics – from sand mining and beach renourishment to sand artistry and displaying your collection. I'm confident that readers will find something for almost any interest in psammophilia.

In addition to attending the 2013 *SandFest* and lecturing on sand to various groups, for the past five years I've had the privilege of editing the *Sand Paper*. That has surely been an education in itself. So how did I become interested in sand collecting? I suppose it all started as I grew up in Reading, Pennsylvania, and became an avid rockhound. Back then, most of my specimens were way too big to qualify as sand. But my interests in materials and microscopy led me to a career in metallurgy and materials science where I eventually specialized in the friction, lubrication, and wear ('tribology'). One need only consider the processes of abrasion and erosion to realize how important a role particles play. As time went by, space for my rock collection became an issue. While I still enjoy collecting mineral specimens, my hobby eventually led me to downsize to micro-mounts, and finally to sand. My sand collection has yet to exceed 1,000 samples, and that is modest compared with the collections of many ISCS members. But I still have millions upon millions of grains to view, photograph, and enjoy.

The more I dug into sand (pun intended) the more fascinating a subject it became, and for the past few years I taught a course called "The Amazing World of Sand – Art, Nature, and Technology" in the Osher Lifelong Learning Institute hosted by the University of North Carolina – Asheville. No doubt, the new information about sand that I gleaned from assembling this symposium will strengthen future presentations of that course.

In the United States and in the wider world, it's a pretty rare thing when so many arenophiles gather to share our interest. I hope this symposium and book of abstracts will spur your interest in sand and in contributing to ISCS in many ways – art, earth science education, and the joys of collecting.

Peter J. Blau
Symposium chair
SandFest 2018

Presentations

Black Gold

Craig B. Hamilton
President, Southeast Federation of Mineralogical Societies, Inc.
Jacksonville, Florida USA

Introduction. Heavy mineral mining began in Northeast Florida in 1913 by mining the black sand located at South Pablo Beach. The beach area mining ended by 1932. Another mining company named Humphrey's Gold Mine started in 1940 in Jacksonville in an area known as east Arlington. Here the firm continued operations until 1957. The E.I. DuPont Company began mining just south of Kingsley Lake at Trail Ridge, about 6 miles east of Starke, Florida in about 1960. That was the beginning of the mining along the Trail Ridge area and it has continued north to just off I-10 near Macclenny, Florida in southeast Baker County. Additional mining was operated to the south of Green Cove Springs, Florida. This presentation focuses on the speaker's past and present experience with heavy mineral mining in the Northeast Florida area.

Mining History

Early 1900s. The beach mining process was begun by Henry H. Buckman and George H. Pritchard and the location was known as "Mineral City". Today it is called Ponte Vedra Beach. Buckman was a graduate of Harvard and Pritchard was an engineer. Mineral City was located about 3 miles south of what is now Jacksonville Beach. This mining began in 1913 and continued until 1932. Buckman had discovered ilmenite (black specks in the sand) which is a titanium mineral ($FeTiO3$). The beach mining was started using mules and carts, but later a rail line was completed in order to move the material for shipping to Jacksonville. The mining company bought 19 miles of beach property south towards St. Augustine for $1,100.00.

There is "Gold" in them there hills. Heavy mineral mining resumed in the east Arlington area in 1940. The mining company was known as Humphrey's Gold Mine. There the company mined along an area of ancient sand dunes from what is now Regency Square area and south along Southside Boulevard for about 3 miles. The company recovered rutile, titanium, ilmenite, and zirconium. Mining continued in this area until 1957.

The E.I. DuPont Company resumed heavy mineral mining at Trail Ridge just south of Kingsley Lake and east of Starke, Florida. DuPont has continued mining along the ridge past Lawtey and today the operation is located in southeast Baker County near I-10.

A material of many uses

The heavy mineral mining industry in Florida provides various work opportunities and raw materials for many types of manufacturing and in numerous products. Titanium is just one of the minerals that has many uses in our everyday lives. It is used in paint, metals, toothpaste, and even as a coating on candy (for example, M&Ms and Skittles) to prevent sticking of the candy before the shell is applied. So: "You like DIRT? Eat Up!"

Restoration

Restoration is a very important responsibility of the overall mining industry. While the yield from heavy mineral recovery pays huge dividends, the land by law must be reclaimed. Not all sand mining lands have been restored, but in recent years upon completion of operations the plan must include reclamation. In simple terms, the land has to be contoured and restored, including water features and planting local indigenous trees and plants.

Summary

As you can imagine from what is outlined here, "Heavy Mineral Mining" production in Northeast Florida is a money maker with large profits. The mining industry has long provided valuable raw materials which are vital to industry and the consumer. Sand mining is for business and the reclaiming program is to restore the land to prior conditions for future use and development.

About the Speaker

A life-long resident of Jacksonville, Florida, Craig Hamilton is one of twenty-five original charter club members of The Jacksonville Gem and Mineral Society (JGMS), Inc. The club was formed in 1959 and he joined in the fall of that year. He served as President in 1971/ 1972, and again in 2016. He currently chairs the JGMS COPS program (Community Outreach Program). This is an educational outreach to public and private schools, the Boy Scouts, and the Girl Scouts. Craig is also responsible for the club's library displays.

After serving in the U.S. Air Force from 1964 thru 1968, Craig's 45-year professional career, included experience was as a board and electronic draftsman and as a mechanical and civil engineer. His involvement in manufacturing and civil engineering projects included many water and sewer projects for streets, highways, and parks. Craig is currently President of the Southeast Federation of Mineralogical Societies, Inc. (2018).

Overview of Heavy Mineral Sands Production*

Jim Renner
Manager, Environmental Stewardship
Southern Ionics Minerals, LLC

Introduction. "Heavy mineral sands" describes sand deposits that contain a higher than usual concentration of minerals that have a specific gravity greater than that of quartz (approximately 2.6 g/cc). Heavy mineral sand deposits are formed when igneous and metamorphic rocks are weathered and eroded. Erosion-resistant minerals (mostly quartz sand) are transported by rivers to the coast, where wave action washes away some of the lighter quartz sand, increasing the proportion of the heavier mineral sands up to as much as 4%. Important heavy mineral sand deposits are found in the United States, Australia, and several African countries.

Heavy minerals of greatest economic interest include zircon, rutile, ilmenite, and leucoxene. Other heavy minerals of economic interest include staurolite, kyanite, and two rare-earth phosphates monazite and xenotime. Gangue (waste) heavy minerals include epidote, magnetite, garnet, tourmaline, and others.

Heavy mineral sands have been mined in the southeastern United States since the early 1900s. Since Iluka's mines in Virginia closed in 2016, Chemours (formerly DuPont) and Southern Ionics Minerals (SIM) are the only domestic heavy mineral sand producers. Chemours continues to mine ancient beach deposits in north Florida on Trail Ridge that DuPont began mining in the 1940s. SIM mines another ancient beach ridge, the Penholoway Terrace, in southeast Georgia.

Southern Ionics heavy mineral sands mining. Southern Ionics Inc. (SII) the parent company of SIM began looking for a domestic zirconium source in 2008. The global supply eventually stabilized, but SII persisted in establishing a zircon sand mining operation. In 2013, permits were obtained for the Mission North and Missing South sites. Construction of the Mission South site in Charlton County, Georgia, began in 2013 and by mid-2014 heavy mineral sand concentrate was being produced. Soon thereafter, SIM constructed a mineral separation plant in Pierce County and began separating the mine-produced mineral concentrate into three main sand products: zircon, rutile, and ilmenite.

The Mission South Mine. The Southern Ionics Mission South Mine (MSM) is located about 50 miles inland from the Atlantic shoreline. Geologically, it is located on the second highest and second oldest coastal terrace: the Penholoway Terrace. The heavy mineral sands concentration ranges from 1.5 to 5% and the deposits extend over roughly 2000 acres to a depth of about 25 feet. Compared with other deposits in Georgia and Florida, the MSM deposit is finer-grained and has more zircon and rutile. In addition, its leucoxene has a higher Ti content.

*This extended abstract was summarized and abridged from the two references listed at the conclusion.

Mining process operations. The following operations take place at MSM. These operations are described in more detail in Ref. [1].

- Trees are harvested, and roots and stumps are removed. Topsoil is stripped and stockpiled for later replacement.
- A rim ditch is created to begin dewatering the next 10 acre area to be excavated.
- An excavator digs the sand ore and loads it into 40-ton haul trucks which take it to the Mining Unit stockpile. (see the figure below)
- The ore is fed into a trommel (rotating screen) to remove oversized material such as woody debris and chunks of humate hardpan (organic-cemented sand).
- Sand falls through the trammel and is mixed with water and pumped to the wet mill. The sand-water slurry travels through numerous sets of spiral flumes where the heavier minerals segregate themselves from the lighter minerals. The heavy mineral content is increased from 2% up to about 80% as the lighter minerals are rejected and returned to fill in a previously mined pit
- Heavy mineral concentrate is pumped to the attritioner that uses grain-to-grain abrasion to remove humate (organic) coating
- After rinsing and stockpiling, the mineral concentrate is trucked to a mineral separation plant in Pierce County (dry mill) for further processing.

An excavator loads ore into a haul truck in the active pit. (Adapted from Fig .4, reference [1].

At the mine site, attention is paid to wetlands protection and avoidance, water reuse and non-discharge of process water, protection and re-establishment of the water table, topsoil replacement, and revegetation, including pine tree replanting.

Heavy mineral sand product usage. There are many industrial uses of zircon and titanium mineral sands. Zircon is an important component of ceramics and is used in glazes like those found in bathroom fixtures. It's also used in molds for casting specialty metal alloys with for aircraft engine parts. Corrosion-resistant Zr alloys are used in cladding for nuclear reactor fuel rods and recently, in dental implants. Ti-containing rutile, ilmenite, and leucoxene are used to make bright white titanium oxide pigment which replaced lead in paint.

Summary. Heavy mineral sands are an important natural resource for northeast Florida and southeast Georgia. The deposits are world-class sources of zircon and ilmenite that are critical minerals for many industries. Advances in mining technology and good environmental management practices will allow these deposits to be responsibly utilized for decades to come.

References

1. James Renner (2016) "Southern Ionics Minerals, LLC Mission South Mine, Charlton County, Georgia," *Georgia Geological Society Guidebook*, Volume 35, pp. 69-74.

2. James Renner (2018) "Field Trip 5: Heavy Minerals Mining, Georgia. *ISCS SandFest 2018 Field Trip Book*, ed. Lori Carter.

About the Speaker

James Renner is the Manager of Environmental Stewardship at Southern Ionics Minerals, LLC, whose corporate office is located in Offerman, Georgia. Southern Ionics Minerals is a subsidiary of Southern Ionics Inc., a chemical manufacturing company with a dozen facilities in the southeastern United States.

Coastal Geology and Beach Nourishment

Jennifer L. Coor, Ph.D., P.G.
Southeastern Geological Society
Jacksonville, Florida

Introduction. The loss of sand from Florida's beaches and coastal systems is a serious problem that affects not only the coastal system, but also the economic livelihood of Florida's coastal communities. Beach erosion is a chronic problem in Florida, influenced by sea-level rise, natural coastal processes, storms, and inlet management [1]. Florida has approximately 1,350 miles of shoreline, from which a variety of different environments are found. Of these, 825 miles are made up of sandy shorelines fronting the Atlantic Ocean, Gulf of Mexico, and Straits of Florida. The Florida Department of Environmental Protection has documented 411.2 miles of critically eroded beaches, 8.7 miles of critically eroded inlet shoreline, 93.5 miles of non-critically eroded beaches, and 3.2 miles of non-critically eroded inlet shoreline statewide in a Critical Erosion Report, as shown in Figure 1 [2]. Of these, 227.4 miles are under active management, having a total of 78 listed projects statewide, as shown in Figure 2 [3].

The beach, aka shoreline, is simply the region where the ocean touches dry land. Florida beaches are commonly composed of sand and shell (carbonate), though in some areas exposed outcrops of rock are visible on the beach (Flagler County and Palm Beach County). The material on the beach is influenced by longshore current, proximity to sediment sources (rivers), land topography and composition, tectonic activity, configuration of the coastline (presence of inlets), wind and weather patterns, among other factors such as sea level rise and anthropogenic activities. Erosion refers to material that is lost from the beach, while accretion refers to the addition of material to the beach.

Beach Nourishment. Florida's eroding beaches first began receiving sand as "beneficial use" from maintenance projects as early as the 1940's, but beach nourishment practices did not begin until the early 1970's. The FDEP created the "Sand Rule," Florida Administrative Code (F.A.C.) 62B-41.007(2)(j)(k), in 1992 and amended the Rule in 2001, which specifies the criteria for sand placed on beaches within the state of Florida. In an effort to protect the environmental function and general character of the coastal system, the "Sand Rule" requires that material used for beach nourishment be similar in grain size distribution to the native or existing beach. Additionally, the "Sand Rule" specifies that fine sediment, defined by the percent of material passing the #230 US Standard Sieve, must comprise 5% or less of beach nourishment material. Material obtained from maintenance of navigation channels allows 10% fines or less for beneficial use on the beach, and 20% fines or less for beneficial use in the nearshore. Identifying offshore sand resources suitable for future beach nourishment projects is critical in order to sustain Florida's beaches.

Northeast Florida. The majority of the beach nourishment projects completed in northeast Florida utilize material obtained from offshore sand sources and the maintenance dredging of navigation channels in both the Intracoastal Waterway (IWW) and coastal inlets. Upland sources are occasionally used for small-scale beach nourishment and/or dune restoration projects.

Central Florida. Beach nourishment projects along the Atlantic central coast utilize a combination of offshore sand sources, sand obtained from maintenance dredging, and upland sources to maintain the beaches.

Southeast Florida. Many beach quality sand resources located adjacent to the southeast Florida beaches have been exhausted [4]. Additionally, there are many sensitive environmental resources, such as coral reefs, that limit dredging activities. In many locations it is becoming a difficult or impossible to find offshore sediment sources that meet the "Sand Rule" for beach nourishment, resulting in the need for truck-haul projects utilizing upland sand sources.

Panhandle / Northwest Florida. The majority of the beach nourishment projects completed in the Florida utilize material obtained from offshore sand sources and the maintenance dredging of coastal inlets, and associated tidal shoals. Upland sources are occasionally used for small-scale beach nourishment and/or dune restoration projects.

Southwest Florida. Beach nourishment projects along Florida's southwest Gulf coast utilize a combination of offshore sand sources, sand obtained from maintenance dredging of inlets, and accompanying tidal shoals, and the Gulf Intracoastal Waterway (IWW), and upland sources to maintain the beaches. Upland sand sources are used for many projects in Collier and Monroe Counties, due to the few offshore sand sources available.

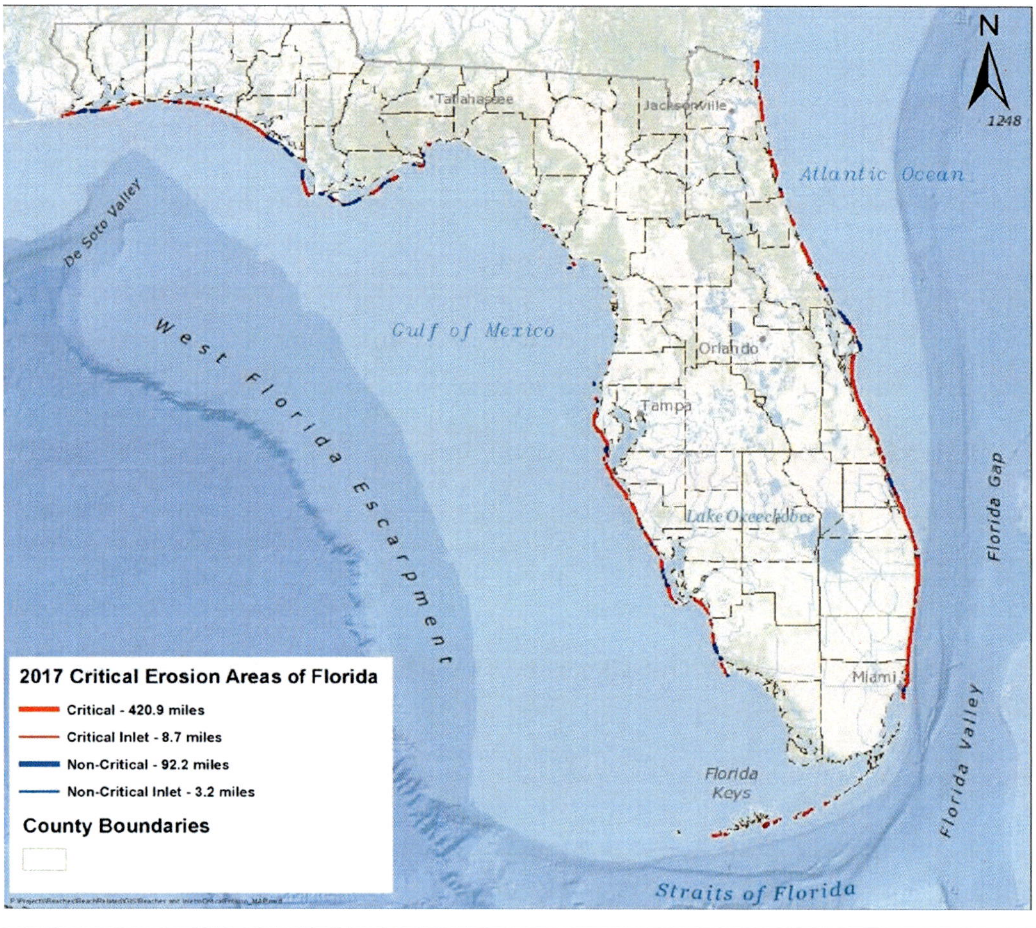

Figure 1. Statewide areas of critically and non-critically eroded shoreline [2].

Figure 2. Florida's managed/engineered beaches (FDEP, 2015).

Conclusions. Many of Florida's beaches are maintained to offset erosion. Beach nourishment occurs throughout Florida's many regions utilizing a variety of sediment sources: offshore sand sources, inlets and tidal shoals, navigation channels, and upland sand sources. The selection of a sand source for a project is dependent on the quality of the sand source and resulting compatibility with the native/existing beach and cost to utilize the sand source.

References. Reference numbers correspond to the bracketed numbers in the text.

1. Clark, R.R. (1993) "Beach conditions in Florida: a statewide inventory and identification of the beach erosion problem areas in Florida," *Beaches and Shores Technical and Design Memorandum no. 89-1*, Florida Department of Environmental Protection, Tallahassee, FL, 202 p.
2. Florida Department of Environmental Protection (2016) "Critically eroded beaches in Florida," Tallahassee, FL, 89 p.
3. Florida Department of Environmental Protection (2015) "Strategic Beach Management Plan," Tallahassee, FL, 337 p.
4. Ousley, J.D., Kromhout, E., Schrader, M.H., Lillycrop, L. (2014) "Southeast Florida sediment assessment and needs determination (SAND) study," *Final Report, ERDC/CHL TR-14-10*, U.S. Army Corps of Engineers, 322 p.

About the Speaker

Jennifer L. Coor, Ph.D., P.G. is a geologist with the US Army Corps of Engineers in Jacksonville, Florida, and was previously employed as a coastal geologist for the FDEP. Jennifer obtained a certification in Coastal Engineering from Old Dominion University in 2017, doctorate in Geology from Florida State University in 2012, and Bachelor of Science degrees in both Chemistry and Marine Science (emphasis in Marine Geology) from Coastal Carolina University in 2006. She has broad research interests including modern and prehistoric storm frequency and impacts, sea level change, geomorphology, sedimentology, geochemistry, and other special projects. She is currently the vice president of SEGS, and helped plan and lead the Honeymoon Island field trip in 2015. Her dissertation focused on determining a storm record for northwestern Florida (panhandle) through a new paleotempestology method, which combined sedimentologic, geochronologic, and geochemical analysis techniques.

Grainatomy: A world in a single grain

Peter J. Blau, PhD, Member ISCS
Blau Tribology Consulting
Biltmore Lake, NC

Introduction. It's been claimed that each grain of sand is unique in the universe. From a strict perspective, that claim is impossible to prove without having compared it to every other grain in the universe. Yet, from a practical standpoint, each grain of sand embodies a host of distinguishing characteristics which, when taken together, differentiates it from practically all others. Collectively, six characteristics of sand grains can be called their *grainatomy*. These are summarized below (see also Figure 1):

1) Coloration
2) Chemical composition and differentiation
3) Size (as part of a larger distribution)
4) Shape (roundness, angularity)
5) Surface micro-scale roughness
6) Microstructure (internal structure)

The term *grainatomy* was coined for a course titled "The Amazing World of Sand: Art, Nature, and Technology" which the author taught in Asheville, NC [1]. The book by Welland [2] was used as the course text, but other references were also recommended [3, 4]. The term *grainatomy* is based on how sand is defined and how the characteristics of individual grains reveal details about their origins, travels, and the processes that created their shapes and surface features.

Billions of Grains. Even a small vial or collecting bag of sand can easily hold millions of grains. Consider for example, a cylindrical vial with an interior diameter of 16 mm and a length of 53 mm long (volume = 10,054 mm^3). Allowing for space between equal-sized, round grains (say, a space-filling factor of 60%), one such vial can contain as many as 1,530 2.0-mm diameter grains, or a whopping 50.1 million 0.062-mm diameter grains! Thus, the grainatomy of sand from different locations presents the collector, geologist, mineralogist, forensic specialist, and sand artist with endless possibilities. Samples collected just a few meters apart on the same beach or on neighboring geological formations, can have much different grainatomies, so the location and date collected should be an essential part of a sand collector's records. This talk will discuss the six characteristics of sand grains. It is not intended to be all inclusive, since that would require a whole book to explain. However, we shall see how some characteristics are more quantitatively measured than others. Interestingly, as will be seen, the process of characterizing and quantifying a sand's grainatomy can present numerous challenges.

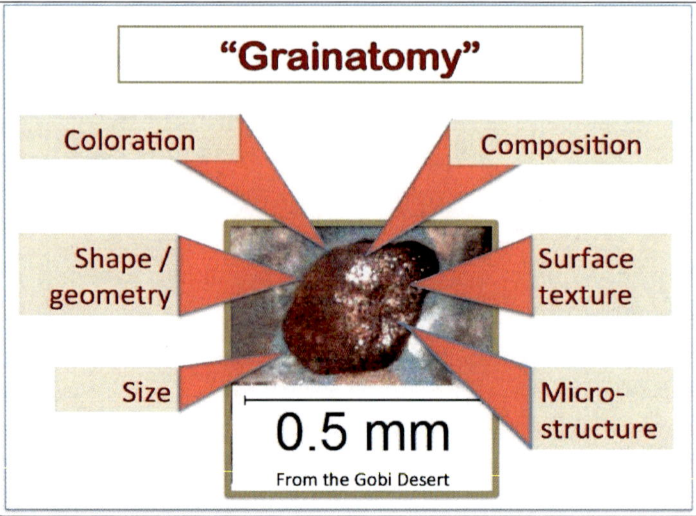

Figure 1. Six characteristics of a grain of sand.

Six Characteristics of Sand Grains.

Color and Composition. Those who collect sand are familiar with its amazing diversity of colors. In turn, those colors depend on composition, particle size, and optical properties. As Welland notes [2], an estimated 70% of all sand is composed of silica (SiO_2). However, to assume that it is present only as quartz would be naïve. Grains can contain heterogeneous mixtures of minerals, chemical species, or biological artifacts. Therefore, the bulk chemical composition of an entire grain can be much different than that of the individual species that comprise it. For example, the Ottawa grains in Figure 2(a) are essentially pure silica, but the grains in Figure 2(b) are heterogeneous mixtures of minerals of volcanic origin. Likewise, a sample of sand can contain grains of different minerals Figure 2(c). Biogenic sands can contain all manner of shells, fossils, and other remnants of living things. Their composition can be silicates, carbonates, and more.

(a) (b) (c)

Figure 2. Sand samples of varying homogeneity. (a) Ottawa foundry sand (nearly pure quartz), (b) Volcanic tephra from Mt. Lassen, California (each grain a mix of minerals), and (c) Mixture of mineral grains from Chunky Gal Mtn., North Carolina (grains of different minerals).

Size and size distribution. As simple as it may first seem, the size of sand grains can be measured in many different ways, and the method of measurement can give different answers. While sand is technically defined as being solid particles ranging in size from 0.063 to 2.0 mm, what do we really mean by "size" ? If all grains were perfect spheres, the answer might be a lot simpler, but rarely are grains perfect spheres. Figure 3 (a) shows some Copacabana Beach sand, some of whose grains are close to spherical, but table salt (b) is cubical, while some sand is very irregular in shape due to its origins by crushing of rock under moving glaciers, biogenic origins [fractured shells or tiny sea creatures (c)], or through other interactions with their surroundings.

Figure 3. Sand samples of varying shape and size. (a) rounded grains from Copacabana Beach, Brazil, (b) cubes and spheres of table salt, and (c) biogenic sand grains in a range of shapes and sizes from Micronesia.

Sand samples containing a uniform distribution of sizes is called 'well-sorted' but sand of a variety of sizes is referred to as "poorly sorted." Defining size in terms of the Feret or caliper diameter (maximum length that fits between two parallel lines), average diameter, or other is affected by how the samples are mounted (for example, loose on a microscope slide, glued to a slide, mounted in epoxy after stirring to ensure randomness and then cross-sectioned). Sizes measured in a microscope and those determined by passing samples through a series of sieves and weighing the retained material on each screen size can easily produce different answers as well. Therefore, the term "size" should not be used alone without defining how it was measured.

Shape. Most grains are not perfect spheres and their shape is one of their most difficult characteristics to quantify. The terms "roundness" or "sphericity" and "angularity" are sometimes used. There are published charts that show increasing roundness in one direction down a column and increasing angularity across a row (for an example, see [5]). The Krumbein chart is another approach to visually depict particle shape [6]. Observing drawings of particles displaying various combinations of roundness or angularity, the user decides which combination best suits his or her sample. Computer programs can be used to capture images of thousands of magnified grains and determine quantities like length, width, aspect ratio (length/width), circularity, perimeter ratio, and other mathematical parameters. These can be related to, for example, the flow characteristics or abrasiveness of a sample of sand grains.

Surface texture. In engineering terms, surface texture can be characterized by *roughness* (fine-scale roughness superimposed on the coarser features), *waviness* (long-period undulations), and *lay* (directionality of features). While a family of precision measuring instruments has been developed to quantify the surface texture of manufactured components like bearings and gears, in the case of sand grains, we might want to depict additional aspects of the surface features of grains that relate to their origins and travels to reach the site where they were collected. The terms listed in Table 1 is one approach to coding grain textures. The terminology listed in Table 1, is not claimed to be a current standard or an all-inclusive scheme, but rather it is proposed here simply for convenience in differentiating surface textures of different kinds of sand grains.

Table 1. Proposed Terms to Characterize Surface Textures of Sand Grains

Term	**Description**	**Code**
Micro-smooth	Shiny, glassy, polished-looking surfaces with no obvious abrasion marks. Characteristic of particles formed by solidification of liquid droplets or by tumbling in water.	M-Sm
Micro-scuffed	Surfaces rounded but not rough. With a glazed, scratched appearance from mechanical tumbling, abrasion, or erosion processes. Similar to those in Figure 2 (a).	M-Sc
Faceted	Containing shiny, regular geometric facets like those of mineral crystals or cubes of salt. Faceted surfaces can eventually become marred or rounded by weathering to blunt their sharper points and edges.	Fac
Fractured	Surfaces displaying thin edges or shiny "fracture mirrors" produced by crushing larger grains. These are more jagged and less geometric than faceted surfaces. Compare for example the scalloped chips from a broken glass with the more regular facets on a crystal of salt. Sand samples containing fractured grains are often poorly sorted.	Frx
Fluffy	A fluffy appearance like that of pollen grains or spiny biological structures. Also, characteristic of volcanic ash or porous pumice from an explosive eruption. Because of their complicated, re-entrant features such surfaces are difficult to characterize.	Flf
Pebbled	A bumpy, granular texture such as that on tumbled composite mixtures like those shown in Figure 2(b), or on the grain of desert sand in Figure 1. Highly pebbled surfaces suggest long periods of exposure to tumbling, flow, and travel from their origins.	Peb
Patterned	Regular patterned features like those on tiny fossils or diatoms of biogenic origin. Examples of patterned features are seen on the shell fragments in Fig. 3(c).	Pat

Microstructure. The term *microstructure* is borrowed from fields such as metallurgy, mineralogy, and ceramics science. It refers to the arrangement, sizes, and compositions of micro-scale features within the interior of a solid material. Examination of microstructure often involves specimen preparation by progressive polishing with finer and finer abrasives, cross-sectioning, and/or chemical etching; however, the microstructures of transparent sand grains can sometimes be examined in transmitted light under a microscope using crossed polarizing filters to reduce glare and reveal internal details. In cases where grains are opaque, dark, or too irregular in shape to use transmitted light, more tedious preparation methods are used. Figure 4 compares transmitted illumination of one type of sand with a polished and etched cross-section of another type of sand (a meteor impact-melted spherule) [6]. Much can be learned from microstructures of grains. They can be probed for chemical analysis, inclusions can be studied to determine the geological ages of the formations from which they came, and the spacing of structures like the dendrites shown in Fig. 4(b) can be used to estimate the cooling rate of a solidifying metal spherule. The degree of porosity inside grains can also indicate its formation conditions.

(a) Black sand from Punalu'u, Hawaii, showing the inclusions inside golden-colored obsidian glass grains. (transmitted light)	(b) Impact melted and re-solidified Fe-Ni-Co-S-P spherule from Canyon Diablo Meteor Crater, Arizona. Diameter ~400 μm. (reflected light)

Figure 4. Comparison of grain microstructures by transmitted light microscopy (a), and by reflected light after mounting, grinding, polishing and etching (b).

Summary. Six characteristics of sand particles, here called *grainatomy*, were presented to illustrate some of the many ways in which the physical features and appearance of sand samples can be differentiated and characterized. The color, composition, size, shape, surface texture, and microstructures of grains can be measured and used in different ways to better understand the origins and properties of sand samples, but the results of those measurements can be affected by the measurement method. While grainatomy goes part way to describing the uniqueness of different kinds of sand, it does not describe their mechanical properties like hardness, compressive strength, frangibility (brittleness), and fracture toughness. Nor does grainatomy attempt to infer thermophysical properties of sand like density, thermal conductivity, electrical conductivity, or melting point. Chemical reactions with the environment can alter sand's surface texture but those effects must be investigated by more specialized methods. There are many ways to differentiate the structure and properties of sand. Grainatomy is but one of many approaches to studying the world grain-by-grain.

References

1 Internet URL: https://olliasheville.com (accessed 12/24/17)
2. M. Welland (2009) Sand: The Never-Ending Story, University of California Press.
3. R. A. Bagnold (2005) The Physics of Blown Sand and Desert Dunes, Dover Pub. New York (orig. pub. 1954).
4. G. Greenberg. C. Kiely, and K. Clover (2015) The Secrets of Sand, Quarto Pub., USA.
5. Wikipedia, URL: https://en.wikipedia.org/wiki/Roundness (geology)
6. W. C. Krumbein (1941) "Measurement and Geological Significance of Shape and Roundness of Sedimentary Particles," *J. of Sedimentary Petrology*, Vol. 11 (2), pp. 64-72.
7. P. J. Blau, H. J. Axon, and J. I. Goldstein (1973) "Investigation of the Canyon Diablo Metallic Spheroids and Their Relationship to the Break-up of the Canyon Diablo Meteorite," *J. of Geophysical Research*, Vol. **78**(2), p. 363.

About the Speaker

Peter J. Blau is a native of Reading, Pennsylvania. He attended Lehigh University (B.S., M.S) and The Ohio State University (PhD) where he majored in metallurgy and materials engineering. He has worked at the U.S. Air Force Materials Laboratory, the National Bureau of Standards (now called the National Institute of Standards and Technology), and Oak Ridge National Laboratory in Tennessee. He currently consults in tribology (friction, wear, and lubrication) and co-edits an international journal called *Wear*. An ISCS member, he has served as editor of *The Sand Paper* since 2013.

Polarized Light Microscopy (PLM) of Sand Mineral Grains

Thomas J Hopen
ISCS President and Microscopy Advisor

Introduction. The presentation that accompanies this extended abstract is intended to serve as a brief introduction to the fundamentals of sand mineral grain identification by transmitted polarized light microscopy (PLM). PLM allows one not only to observe the physical characteristics of a mineral grain but also to determine a variety of optical properties. Basic PLM identification characteristics are shape, color (pleaochroism), refractive index (indices), and birefringence.

In addition to PLM, other techniques such as stereomicroscopy (SM) and scanning electron microscopy with energy dispersive spectroscopy (SEM-EDS) will be briefly discussed. SM is the main tool for the identification of biogenic sand grains because one can observe and determine their grain size, morphology, surface texture, and color. By using the powerful combination SEM-EDS, not only can one study the surface textures and surface inclusions in more detail, but also determine the elemental composition of the grains.

Introduction to PLM. As with most analytical techniques, some sample preparation may be required. It may be as simple as isolating a particle of interest or mounting a well-sorted sample of fine sand. Alternatively, you may need to sieve the sand sample into individual fractions and/or do a density separation of the mineral grains. A density separation allows one to do a more detailed examination of the heavy fraction minerals (i.e., those with a density >2.89 g/mL).

For reference, the specific gravities of some common minerals are listed in Figure 1. The method used to separate them in this case was to add a prepared sand sample to a liquid called 'bromoform'. Bromoform has a density of about 2.84 g/mL so denser minerals will sink and less dense minerals will float when a sample is mixed into the fluid. Some particles may have densities relatively close to that of the bromoform and therefore they may either sink or float.

Design and use of a polarized light microscope. The design of a polarized microscope differs slightly from that of a biological microscope. Its additional features usually include a polarizer located in the condenser section, a circular stage, a slot to insert compensators into the body tube, an analyzer (a second polarizer) which can be inserted as needed, and a Bertrand lens. A more detailed list of mineral characteristics that can be determined by using PLM is provided in Figure 2.

Using a single polarizer, compared to looking at the sample with no polarizer, certain additional characteristics can be determined. These characteristics include noting the relief (contrast), color (pleaochroism), and estimating/determining the refractive index (indices). The latter process involves observing a particle as you rotate the stage. When the analyzer (i.e., the polarizer located above the sample whose privilege direction is perpendicular the polarizer in the condenser) is inserted, the particles' retardation colors can be observed and the birefringence can be estimated/determined.

Float on Bromoform

Quartz (2.65) Feldspars (2.5-2.8)
Calcite (2.7) Gypsum (2.3)

May Sink in Bromoform

Micas (2.7-3.4) Dolomite (2.8-2.9)

Sink in Bromoform

Amphiboles (2.9-3.5)	**Limonite (3.8)**
Apatite (3.2)	**Olivine (3.3-3.4)**
Chlorite (2.6-3.0)	**Pyroxenes (3.3-3.6)**
Corundum ((3.9-4.1)	**Rutile (4.2-4.3)**
Epidote (3.2-3.5)	**Sillimanite (3.2)**
Garnet (3.8)	**Straurolite (3.6-3.8)**
Hematite (5.2)	**Tourmaline (2.9-3.2)**
Kyanite (3.6)	**Zircon (4.7)**

Figure 1. Specific gravity of some common minerals in units of (g/mL).

Size Morphology Surface Texture

Crystal System - Refractive Index (indices)

Color – Pleochroism Relief – Contrast

Birefringence Extinction Twinning Sign of Elongation

Interference Figure Optic Sign

Inclusions Alteration - Zoning Microchemical Test(s)

Figure 2. Mineral Characterization by PLM

A chart slowing the relationship between retardation and birefringence is provided in Figure 3. Michel-Levy charts may be obtained from a variety of sources on the internet and in microscopy texts. The particles' extinction can be determined as well as signs of twining and elongation, if the particles possesses those characteristics. Furthermore, if one inserts a Bertrand lens (for conoscopic viewing) the particle may show an interference figure, and the optic sign can be quickly determined. Identification of minerals by PLM is illustrated in Table 1 that follows.

Conclusions. Polarized light microscopy is a powerful tool for the identification of minerals found in sand samples.

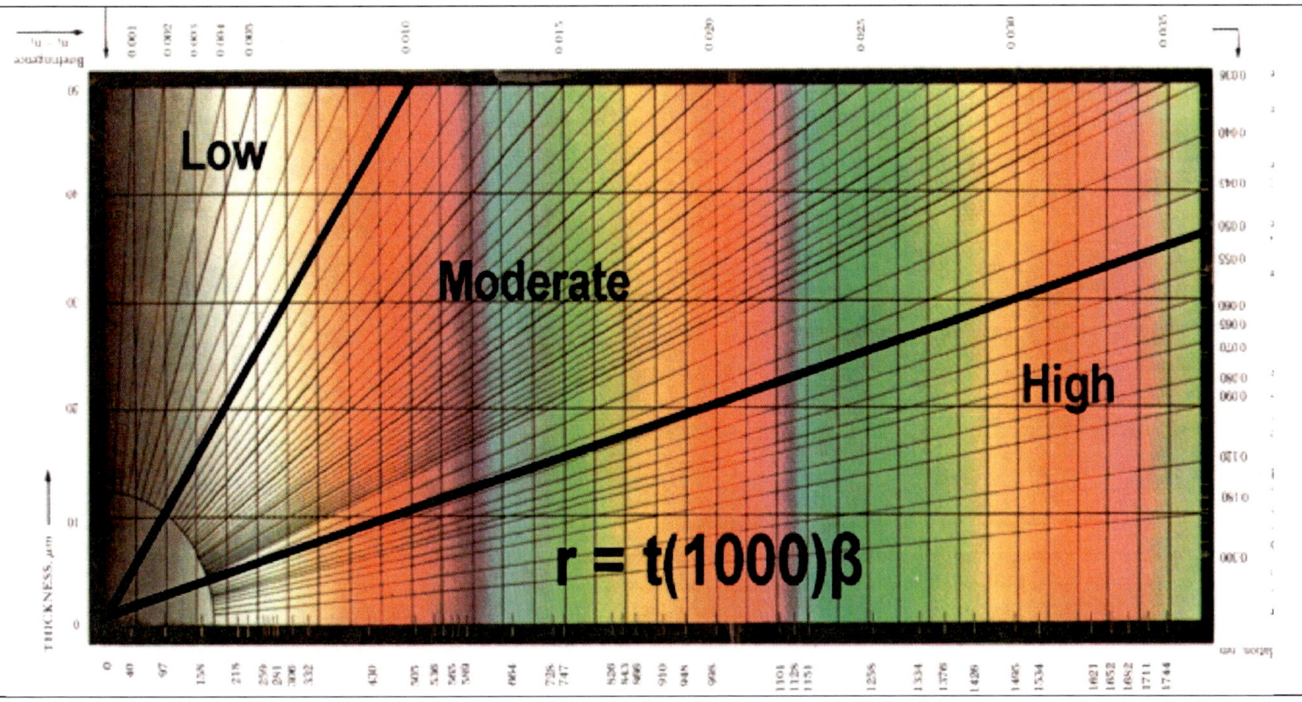

Figure 3. Michel-Levy Chart showing the relationship between birefringence and retardation.

References

Bloss F. An introduction to the methods of optical crystallography, Mineralogical Society of America. Washington DC, 1999.

Heinrich, E. Wm., Microscopic identification of minerals, McGraw Hill Book Company, 1965

Hopen, T. "The value of soil evidence", in M. Houck ed. Trace evidence analysis - more cases in mute witness, Elsevier Academic Press, 2003, pp 105 - 122

McCrone W, McCrone L, and Delly J., Polarized light microscopy. McCrone Research Institute, Chicago, 1978.

Nikon [2007] *Nikon microscopy: Introduction to polarized microscopy*. [Online]. Available from URL: http://www.microscopyu.com/articles/polarized/polarizedintro.html

About the Speaker

Thom Hopen was first introduced to sand collecting by Dr. McCrone in the late 70's and never stopped collecting and studying sand. Thom has over 40 years as a forensic microscopist and was an instructor for several years at the McCrone Research Institute in the 1980's. He received the Émile M. Chamot Award from the State Microscopical Society of Illinois and the Lockard Award from American Society of Trace Evidence Examiners. He retired from the ATF, Forensics Sciences Laboratory, Arson and Explosive Section, Atlanta, GA in 2015.

PLM Mineral Grain ID Reference Table

Mineral	Shape	Color/ Pleochroism	Contrast n vs. medium	Birefrin-gence	Extinction	Other (inclusions, surface features, interfere figure, optic sign, etc.) & additional (*) comments.
Light Fraction Mounting Medium: 1.545 HD						
Quartz	Angular Grains*	Colorless	Low Moderate n's - 1.544 & 1.553	Low	Complete	May have rutile needle inclusions. Conchoidal fractures common. Uniaxial (+). *Grains can be "rounded".
Alkali Feldspar	Tablets*	Colorless	Low n's < 1.540	Low	May show twinning	Biaxial *Usually show good cleavage
Plagioclase Feldspar	Tablets*	Colorless	Very low to low n's ~ 1.53 to 1.59	Low	May show twinning	Biaxial *Usually show good cleavage
Plant Opal	Varies	Colorless	n ~ 1.47	none	NA	
Calcite/ Dolomite	*	Colorless	Varies to v. high 1.48/1.50 to1.66/1.68	High to very high	Symmetrical	*Euhedral rhombs and irregular fractured grains. Uniaxial (-). Calcite normally in light fraction-dolomite normally in heavy fraction
Heavy Fraction Mounting Medium: 1.660						
Augite	Elongated worn*	Colorless to green with no pleochroism	Moderate n's - 1.66 to 1.76	Moderate-High	∠ > 30°	Clinopyroxene – Biaxial (monoclinic). May show optic axis figure. Show blue edges & may have inclusions * Also poorly irregular to rounded.
Biotite	Rounded Flakes	Brown Weak	Low β & γ = 1.57 to 1.68	Low	Complete	Biaxial – 2V + 0 to 25° (-) obtained on most plates. May look uniaxial where 2V is very small. β & γ in plane of flake.
Chlorite	Rounded Flakes	Green	Varies n - 1.57 to 1.67	Low	Complete	Small 2V- usually (+) obtained on most plates. May look like biotite but (+) optic sign.
Epidote	Rounded	Faint yellow to green/yellow	Moderate n's > 1.7	Moderate*	Complete	Biaxial, 2V = 64-89° (-). *May show anomalous interference colors.
Garnet	Rounded grains*	Colorless, pink, salmon & orange	Moderate to high n's – 1.72 to 189	-	-	Green is usually spinal and easy confused as being a garnet. *Also, may see angular irregular grains.
Glaucophane	Elongated grains	Distinctive blue and violet colors	Low n's ≤ 1.67	Low/Moderate	Not ǁ < 15°	Amphibole Mineral
Hornblende	Elongated*	Olive to blue-green	Low n's sl. < & sl. > 1.66	Low	∠ < 25°	Amphibole mineral *Also angular to rounded.
Hypersthene	Elongated grains*	Pink/red to green	Moderate Show blue edges	Low	Parallel	Orthopyroxene - Biaxial (orthorhombic) * Ends may be pointed. Also stubby cleavage fragments.
Kyanite	Elongated /Bladed	Colorless	Moderate n's 1.71 to 1.73	Low/Moderate	Inclined	Triclinic. Grains my show conspicuous cross fractures.
Muscovite	Flakes	Colorless	Low β & γ = 1.55 to 1.62	Low	Complete	Biaxial – 2V = 30-47° (-) obtained on most plates. β & γ in plane of flake.
Olivine	Irregular grains	Colorless	Low n's - 1.6 to 1.9	Low to Moderate	Complete	Orthorhombic. Show brilliant interference colors. Rare as soil mineral but found in sand
Rutile	Irregular elongate*	Red to yellow color Faint to Strong	Very high n's – 2.6 to 2.9	High	Elongated Grains - ǁ	Uniaxial (+) *Also rounded grains.
Sillimanite	Elongated grains	Colorless to rose Weak	Very low to low n's 1.65 to 1.68	Low	Elongated Grains - ǁ	Orthorhombic
Sphene (Titanite)	Irregular grains	Yellow to brown Weak	Very high n's 1.9-2.0	Very High	Incomplete Due to Dispersion	Biaxial. 2V = 27° (+)
Staurolite	*	Colorless to brn/blk. Pleochroic in shades of yellow.	Moderate n's – 1.74 to 1.76	Moderate	Complete	Orthorhombic. Usually has inclusions. *Irregular to somewhat platy and rounded.
Tourmaline	Elongated to round	Colorless to brn/blk. Strong	V. low to moderate n's – 1.62 to 1.69	Low	Elongated Grains - ǁ	Some show blue pleochroism colors. (-) sign of elongation. Tetragonal -Uniaxial (-) on particles that show color in all orientations.
Zircon	Euhedral Elongate*	Colorless	High n's - 1.92 to 2.0	High	Elongated Grains - ǁ	Tetragonal *Also, elongated rounded.

Iron Oxides by reflected light: **Magnetite** - black; **Hematite** - usually black or red; **Limonite** - yellow and is usually a coating on other minerals.

T. Hopen - 02/2012

Table 1. PLM Identification Characteristics of Common Minerals

Characteristics of Florida Sands

Jennifer L. Coor, Ph.D., P.G.
Southeastern Geological Society
Jacksonville, Florida

Introduction. Beaches are dynamic land forms at the edge of the ocean or gulf subject to both natural and man-induced erosion. Sand moves along the shore due to wind driven currents and tides, and storms can cause dramatic changes to the beach. The majority of man-induced erosion is attributed to the creation and maintenance of inlets, where the sand has historically been removed from the coastal system, and the natural drift of sand along the shore is blocked by jetties, trapped in channels, or moved into ebb and flood shoals. The development and the placement of infrastructure in close proximity to the shore has also contributed to coastal erosion by limiting the amount of sand stored in dunes and hardening the shore for protection of upland property. Even on the calmest of weather days in Florida, the sand is always moving within the littoral zone.

Florida has approximately 1,350 miles of shoreline, from which a variety of different environments are found. Of these, 825 miles are made up of sandy shorelines fronting the Atlantic Ocean, Gulf of Mexico, and Straits of Florida. Florida's beaches are one of the State's most invaluable assets, as they serve several important functions, each being vital to maintaining the health of Florida's economy and environment. Florida's beaches are made up of sands that are primarily composed of quartz and carbonate (though the appearance of these sands vary greatly throughout the state.

The loss of sand from Florida's beaches and coastal systems is a serious problem that affects not only the coastal system, but also the economic livelihood of Florida's coastal communities. Beach erosion is a chronic problem in Florida, influenced by sea-level rise, natural coastal processes, storms, and inlet management. Florida has 411.2 miles of critically eroded beaches, 8.7 miles of critically eroded inlet shoreline, 93.5 miles of non-critically eroded beaches, and 3.2 miles of non-critically eroded inlet shoreline statewide [1]. Recent study has shown that beach quality sand resources for restoration adjacent to the beaches along southeast Florida have been exhausted [2]. Identifying offshore sand resources that are compatible with the native and existing beach sediments for future beach restoration projects is critical in order to sustain Florida's beaches that are vital to the economic stability of the state. Florida's beaches are commonly separated into 7 geographic regions, as shown in Figure 1; these designations will be used to characterize the sediments found within each region.

Atlantic Coastline. The beaches of Florida's Atlantic coast can be separated into three (3) segments: Northeast Florida beaches, Central Atlantic Florida beaches, and Southeast Florida beaches. The Northeast Florida region covers the beaches from the Florida-Georgia State line (Nassau County) south through the beaches of Volusia County. The Central Atlantic region covers from the False Cape (aka Cape Canaveral) in Brevard County south to Martin County. The Southeast Florida region covers the beaches of Palm Beach, Broward, and Miami-Dade Counties.

Northeast Florida Beaches. The sediments that make up the beaches of northeast Florida are typically tan to light gray in color and fine- to medium-grained quartz and shell (carbonate) material ranging in size from sand to large gravel. Some beaches have deposits of heavy minerals (Duval and St. Johns Counties), and others are known for coquina rock outcrops (Flagler County). A micrograph of a sample representative of Northeast Florida, collected from Flagler County, is shown in Figure 2.

Central Atlantic Florida Beaches. The sediments that make up the beaches of the central Atlantic coast are typically light brown to gray in color, and are composed of a combination of fine- to medium-grained quartz, shell (carbonate) material ranging in size from sand to large gravel, and small rock fragments. A micrograph of a sample representative of Central Florida, collected from Martin County, is shown in Figure 3.

Southeast Florida Beaches. The sediments that make up the beaches of the southeast Atlantic coast are typically tan to light gray in color, and are composed of a combination of fine- to medium-grained quartz – commonly from upland sources, shell (carbonate) material ranging in size from sand to large gravel, and small rock fragments – some of which are either coral or limestone. A micrograph of a sample representative of southeast Florida, collected from Broward County, is shown in Figure 4.

Northwest Florida. The beaches of northwest Florida can be separated into two (2) segments: Florida Panhandle beaches and Big Bend beaches. The beaches of northwest Florida include the Florida Panhandle (Escambia County to Wakulla County) and the Big Bend (Jefferson County to Pasco County).

Florida Panhandle Beaches. The beaches of the Florida Panhandle are known throughout the world for their famous sugary appearance, gaining them the nickname of "sugar sand" beaches. The sediments that make up the beaches of the Panhandle are light gray to white in color, and are composed of fine- to medium-grained quartz eroded from the Appalachian Mountains. Some locations have trace amounts of heavy minerals, and varying amounts of sand to gravel sized shell (whole and fragments). A micrograph of a sample representative of the Panhandle, collected from Bay County, is shown in Figure 5.

Big Bend Beaches. The sediments that make up the beaches of the Panhandle are light gray to tan in color, and are composed of fine- to medium-grained quartz, shell (carbonate) material ranging in size from sand to large gravel, and small rock fragments. Due to the low energy environment in this portions of the coast, some locations have organic material present on the beach. A micrograph of a sample representative of the Big Bend, collected from Taylor County, is shown in Figure 6.

Southwest Florida. Southwest Florida describes all of the Gulf of Mexico coastal counties from Pinellas County / Hillsborough County to Collier County. The sediments that make up the beaches of the southwest Atlantic coast are typically tan to light gray to white in color, and are composed of a combination of fine- to medium-grained quartz, shell (carbonate) material ranging in size from sand to large gravel, and small rock fragments. A micrograph of a sample representative of southwest Florida, collected from Charlotte County, is shown in Figure 7.

Florida Keys. The Florida Keys describes the chain of islands that makes up the majority of the sandy shoreline in Monroe County. The sediments that make up the beaches of the Florida Keys are typically light gray to white in color, and are composed of a combination of fine- to medium-grained quartz, shell (carbonate) material ranging in size from sand to large gravel, and small rock fragments – commonly either pieces of limestone or coral. A micrograph of a sample representative of the Florida Keys, collected from Key West, is shown in Figure 8.

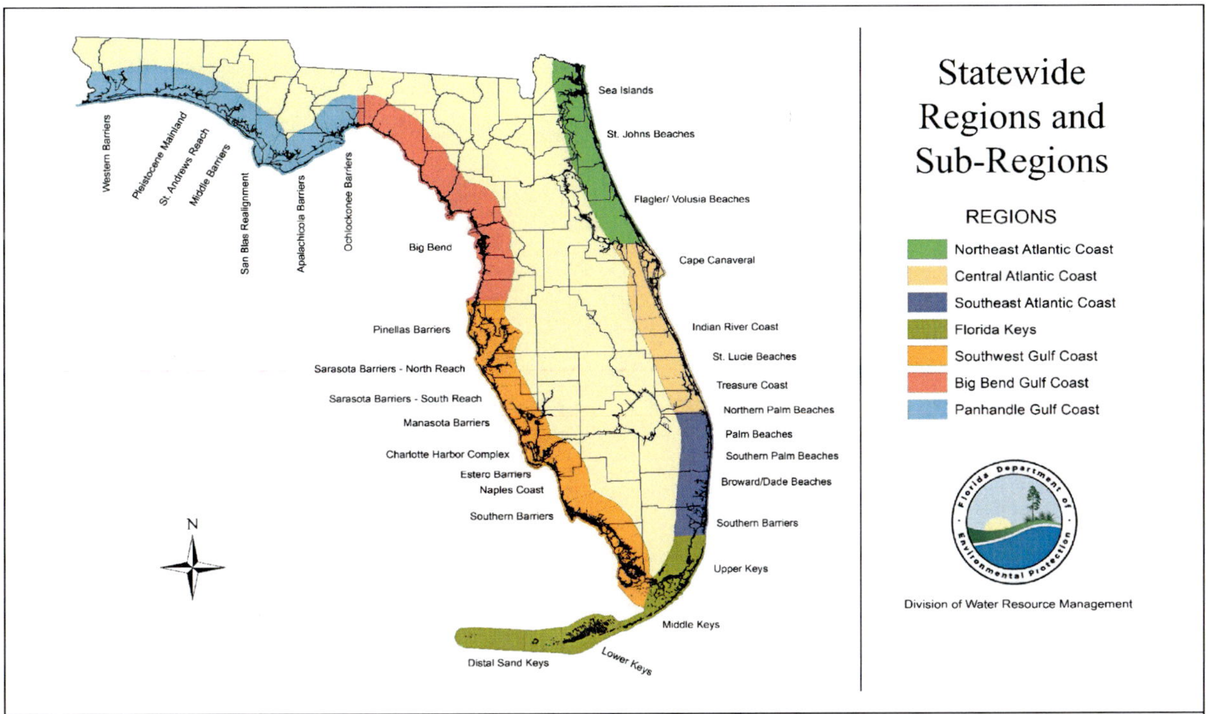

Figure 1. Statewide regions and sub-regions map. [3]

Figure 2. Micrograph for typical sandy beach in northeast Florida. This sample was collected from Flagler County and shows sand sized shell and quartz. Amounts of shell (carbonate) and heavy minerals vary by location. [4]

Figure 3. Micrograph for typical sandy beach in central Atlantic Florida. This sample was collected from Martin County and shows sand composed of darker shell, quartz, and other small rock fragments. Amounts of shell (carbonate) minerals vary by location. [4]

Figure 4. Micrograph for typical sandy beach in southeast Florida. This sample was collected in Broward County and is mostly composed of quartz and small rock fragments. Amounts of shell (carbonate) vary by location. [4]

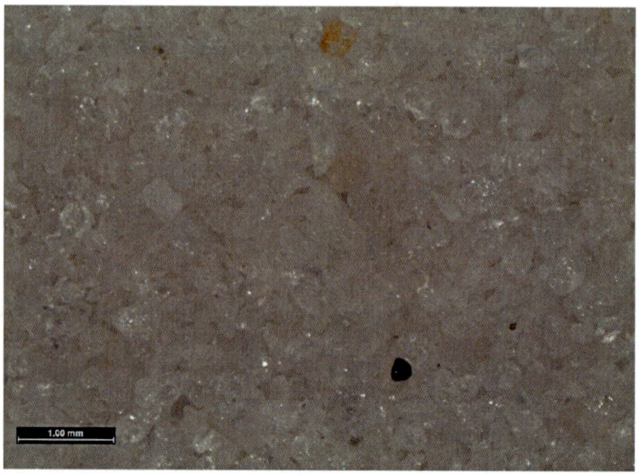

Figure 5. Micrograph for typical sandy beach in northwest Florida. This sample was collected from Bay County, and is composed of fine grained quartz grains with trace heavy minerals. Amounts of shell (carbonate) and heavy minerals vary by location. [5]

Figure 6. Micrograph for typical sandy beach in Florida's Big Bend. This sample was collected from Taylor County, and is composed of quartz, shell fragments, foraminifera, and organic material. Amounts of shell (carbonate) and heavy minerals vary by location. [5]

Figure 7. Micrograph for typical sandy beach in southwest Florida. This sample was collected in Charlotte County and is mostly composed of quartz and small shell fragments. Amounts of shell (carbonate) and rock fragments vary by location [6]

Figure 8. Micrograph for typical sandy beach in the Florida Keys. This sample was collected from Key West in Monroe County, and is composed of a combination of upland quartz sand and carbonate / limestone fragments. Amounts of shell (carbonate) and limestone and/or coral fragments vary by location. [6]

Conclusions. While Florida's beaches are nearly all composed of the same materials – quartz and carbonate – nearly every beach in Florida appears different. Sand sized quartz and carbonate from shell (whole shells and fragments), limestone, and corals, combine in varying proportions; however, the visual appearance of these components changes due to local influences, such as the presence heavy minerals in the northeast and northwest coasts, sediment supply, water quality, size and angularity of the sand grains, and sand source for beach nourishment projects.

References. Reference numbers correspond to the bracketed numbers in the text.

1. Florida Department of Environmental Protection (2016) "Critically eroded beaches in Florida," Tallahassee, FL, 89 p.
2. Ousley, J.D., Kromhout, E., Schrader, M.H., Lillycrop, L. (2014) "Southeast Florida sediment assessment and needs determination (SAND) study," *Final Report, ERDC/CHL TR-14-10*, U.S. Army Corps of Engineers, 322 p.
3. Florida Department of Environmental Protection (2015) "Strategic Beach Management Plan," Tallahassee, FL, 337 p.
4. Phelps, D.C., Ladle, M.M.L., Dabous, A.A. (2009) "A Sedimentological and Granulometric Atlas of the Beach Sediments of Florida's East Coast," Florida Geological Survey, Tallahassee, FL, 35 p.
5. Phelps, D.C., Ladle, M.M.L., Dabous, Harrington, E.P. (2011) "A Sedimentological and Granulometric Atlas of the Beach Sediments of Florida's Northwest Coast and Big Bend," Florida Geological Survey, Tallahassee, FL, 35 p.
6. Phelps, D.C., Ladle, M.M.L., Dabous, A.A. (2010) "A Sedimentological and Granulometric Atlas of the Beach Sediments of Florida's Southwest Coast and Keys," Florida Geological Survey, Tallahassee, FL, 50 p.

About the Speaker

Jennifer L. Coor, Ph.D., P.G. is a geologist with the U.S. Army Corps of Engineers in Jacksonville, Florida, and was previously employed as a coastal geologist for the FDEP. Jennifer obtained a certification in Coastal Engineering from Old Dominion University in 2017, doctorate in Geology from Florida State University in 2012, and Bachelor of Science degrees in both Chemistry and Marine Science (emphasis in Marine Geology) from Coastal Carolina University in 2006. She has broad research interests including modern and prehistoric storm frequency and impacts, sea level change, geomorphology, sedimentology, geochemistry, and other special projects. She is currently the vice president of SEGS, and helped plan and lead the Honeymoon Island field trip in 2015. Her dissertation focused on determining a storm record for northwestern Florida (panhandle) through a new paleotempestology method, which combined sedimentologic, geochronologic, and geochemical analysis techniques.

Why Collect European Sand?

(A pictorial overview of European sands)

Ralf Hermann
Director, Deutches Sandmuseum
Cramma, Germany

Introduction. *Why collect European sand?* – Because of its colors and diversity! Figure 1 compares the land area of the 47 European nations with that for the 49 contiguous states of the U.S.A. In Europe, many sand collectors prefer collecting samples from beaches, and Europe offers about 117,000 km (72,700 mi) of coastline to explore. Greece owns 82% of the islands in the Mediterranean Sea, and they consist of about 3,054 isles, islets, large rocks and reefs. Add to that thousands more islands in the British Isles with their Hebrides and Norway with its Lofoten Islands. However, many of these do not have collectable sand.

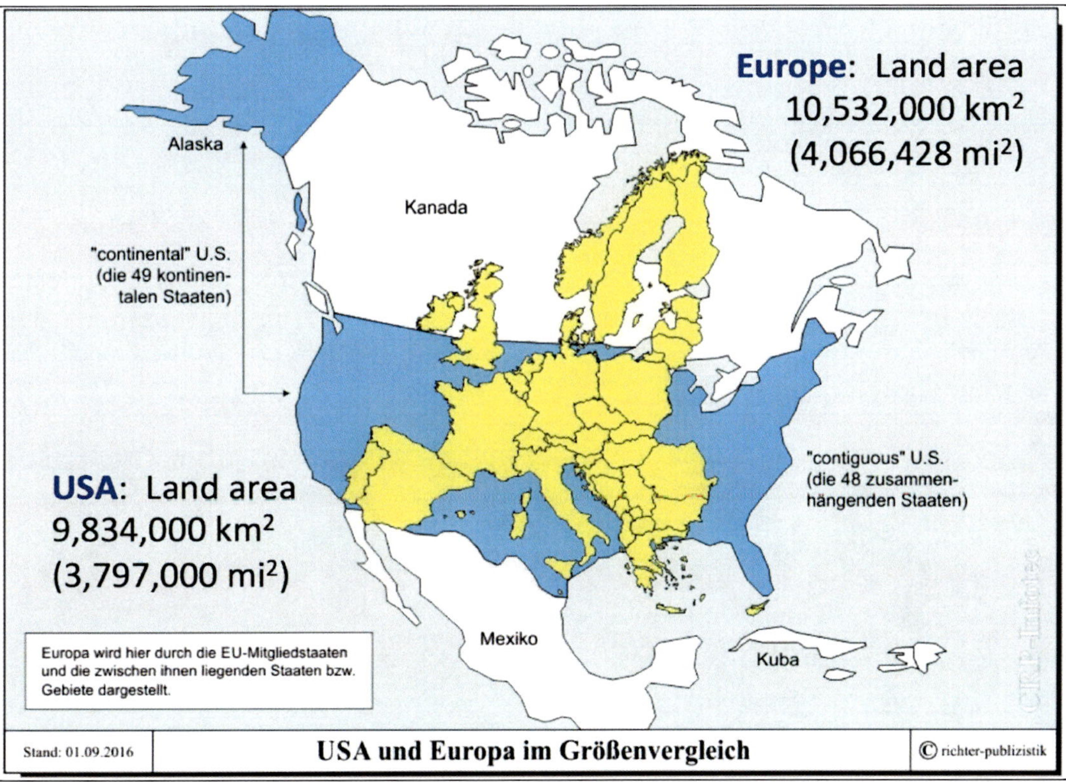

Figure 1. Comparison of the relative sizes of Europe (yellow) and the U.S. (blue) in North America. (Adapted from Richter-Publizistik)

The presence of large and small islands adds to the enormous coastline of Europe. Some large areas have a similar geological history, and the sand samples you find appear the same or originate from the same base material. In this talk we organize different areas along the coastline according to their fundamental primary colors. The presentation identifies different regions of Europe and reveals the diverse, colorful sands that can be found there. Figure 2 shows the regions from which these examples are taken.

Figure 2. Regions of Europe identified by the fundamental primary colors of their sands.

Colors of European sands. The following images show the diversity of colorful sands collected from different regions of the European coastline and along its island beaches. The first example is heavy mineral sand. It contains a mixture of garnets, magnetite and hematite and comes from coastal regions along the North Sea and the Atlantic Ocean.

Figure 3. *Denmark* – Cladved. Heavy mineral sample with garnet, magnetite and spherules (balls).

Figure 4. *Netherland* – Brule. Typical sample of heavy minerals from the North- and Eastsea.

The southern coast and islands of the Mediteranean Sea ("Mittlemeer" in German) offer red forams. These are not overly common except in certain rich deposits. Forams of Genus *Miniacina miniacea* are found near the Mediterranean Sea (Fig. 5). More examples of the genus from Crete are shown in Fig. 6.

Figure 5. Biogenic sands from southern Europe:

Upper left: Italy – Burdelli Island, Spaggia Rosa.

Upper right: Spain – Menorca-Platia de Son Saura.

Lower right: France – Corsica, Cape Corse.

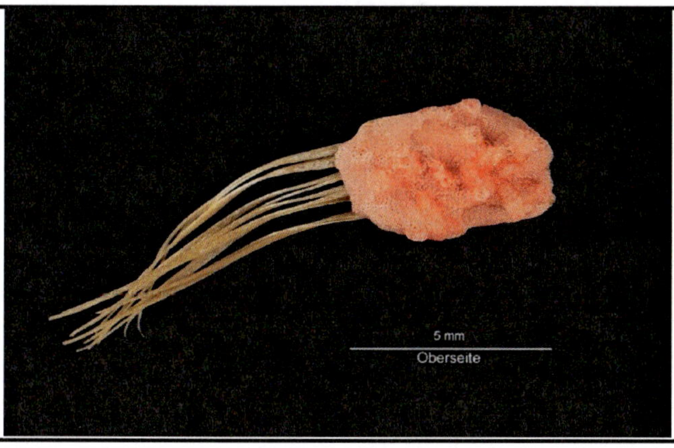

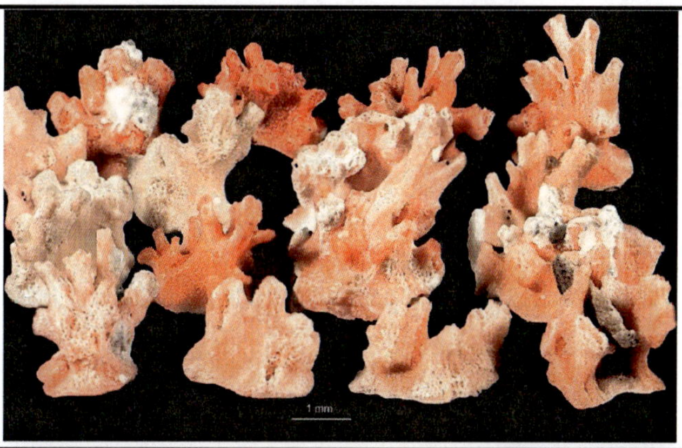

Figure 6. Red forams *Miniacina miniacea* are a bit bizzare but interesting for viewing and very photogenic.

Upper left: Greece, attached to a plant from Crete (Elafonisse Beach and Gramvousa Peninsula.

Upper and lower right: Island of Rhodes, Prasonini Beach

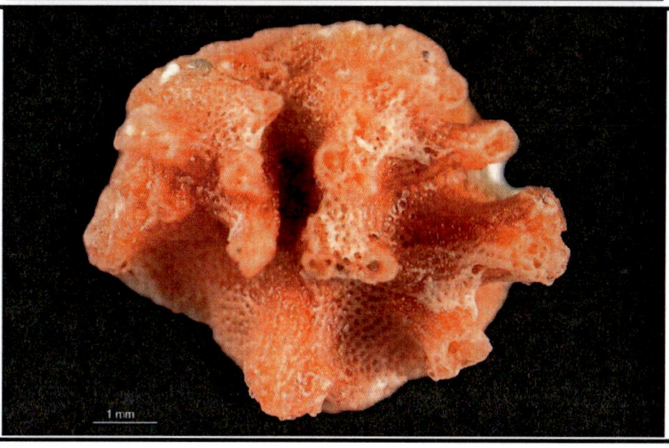

Spain's Canary Islands in the Atlantic Ocean offer a variety of colorful sand samples (see Fig. 7 – 10).

Figure 7. Sample of a Hauyn (tectosilicate mineral) separation from Playa del Camison on Tenerife

Figure 8. Carbonate sand from Lanzerote-Bajo.

Figure 9. Olivine sand from Playa Taurito on Grand Canary Island.

Figure 10. Volcanic sand from Fuerteventura- Puerto Naos

As shown below, other equally unusual and interesting sand samples can be found throughout Europe. Some come from beaches (Fig. 11 – 14) but others come from pits, rivers, and mine dumps (Fig. 15 – 17). Further interesting biogenic and abiogenic (mineral) examples are shown in Figures 18 – 20.

Germany's Salzgitter/Cramme region, birthplace and home of the presenter, offers some interesting sands as well. They include oolitic iron (Fig. 21), other spherules of iron ('ferro-balls') of uncertain origin (Fig. 22), and a Hauyn separation that is also known for its heavy minerals (Fig. 23).

Finally, there is an area of the Harz Mountains that is known for its many interesting kinds of sand. For example, the "Bode Achat" (copper slag) and a green garnet variety called opazolith are two of the nicer mineral grains to be found in samples from this region. Figures 24 – 27 show a few examples.

Conclusions. Europe offers a fascinating variety of collectable sands – like European star stand (Fig. 28). They can be found from the continental coast to the many islands and from the beaches to the interior regions. We welcome visits from fellow psammophiles to collect and trade sand samples.

Figure 11. Endemic Maerlsand from Coral Beach, Ireland

Figure 12. Scheveningen, The Netherlands

Figure 13. Big forams from Pernera Beach, Cyprus.

Figure 14. Oolites from Kleopatra Beach, Turkey.

Figure 15. Sapphire/garnet sand found between the towns of Brettignoles-Sur-Mer and Sable d'Olonne in France. Many places on Island Groix also offer nice, heavy mineral samples.

Figure 16. This endemic mineral called naujakasite is found only in Kvanefield, Greenland.

Figure 17. Faceted pyrite sand is found close to the beaches on the island of Elba in Italy.

Figure 18. Sand containing yellow mollusks with forams and sea urchins from Galway Beach, Ireland,

Figure 19. Biogenic sand sample with a lot of mica from Saint Malo, France, a place which is also known for its maerlsand.

Figure 20. Sand containing some nice, intact forams from Mellieha Beach, Malta.

Figure 21. Oolitic iron spherules from Salzgitter, Germany.

Figure 22. Possible oolitic iron spherules or emission products of unclear origin found in Meesche brook in Cramme, Germany.

Figure 23. Grains of sand extracted from Hauyn in Cramme, Germany. (Same location as Fig. 22)

Figure 24. This rare mineral nontrinite is from the Roter Bär (red bear) Mine in St. Andreasberg, Germany. (see also Fig. 27)

Figure 25. Black glass head (var. Romanechit) from Ilfeld, Harz, Germany.

Figure 26. Red glass head a variety of hematite called specularite, from Ilfeld, Harz, Germany.

Figure 27. Blue lazurite is also from the Roter Bär (red bear) Mine in St. Andreasberg, Germany. It was an industrial product about 1900.

Figure 28. European "Star Sand" is of the genus "Siderolites" and is found in the Rooth sand pit in Maastricht, in The Netherlands.

About the Speaker

Ralf Hermann is a native of the Salzgitter, which is about 75 km southeast of Hanover, Germany. He founded the German Sand Association (German Sandmuseum e. V.) in 1997. This small association prepared an exhibition called Sandwelten ("World of Sand") in Cramme, which is a part of Wolfenbüttel near Braunschweig. There,'everything worth knowing about sand' is on display. In particular, it shows the history of culture related to sand (e.g., the sand industry and its products, biology of sand (animals and plants), the influence of sand in everyday life, sand art, and hourglasses. Of course, it also includes the genesis, types, and collecting of sand from all over the world. All topics related to psammology are being addressed within a continuing range of projects.

At this writing, there are several active projects that would welcome participation and expressions of interest from the reader. These include an activity on the history and technology of hourglasses. In the late 1980s, Hubert Groß (DE) and Raymone Maylers started a project called "Sand and Names". References and source materials were obtained some years ago, and work on a revision of this book has been underway since the late 1990s. Participants are collecting pictures and information about places that contain the word "sand" in their title. That includes all languages and such things as pictures of city signs, company signs, beach signs – in other words, everywhere you can find the word SAND!

Origins of Hermann's interests that resulted in the eventual establishment of the Association and Sandmuseum originated in the late 1980's. Hermann's sand collection increased from 5,000 samples in 1999 to over 45,000 today. A chronology of its history and the founding of its projects may be found (in German) on the website of the museum is www.sandmuseum.de. The museum's e-mail address is: info@sandmuseum.de

Playing in the Sands—A Photographic Tour

Leo Kenney and Kate Clover
ISCS Members

Introduction. The *Playing in the Sands* educational sand poster was first published in 2005 and has been popular amongst sand collectors and educators. A completely new poster is being introduced at *SandFest 2018* featuring beautiful photographs of 50 sands from around the world.

History. The original poster was developed and published less than a year after Leo Kenney began to photograph sands. Distribution inspired several educators to cooperate in later years in producing the *Splendid Sands Calendar*. Kate Clover and Carol Hopper-Brill continue to collaborate with Kenney on the sand calendar and now the new *Playing in the Sands* Poster. Twelve years of collaboration and continued feedback from users brings new photography, text, and theme to the poster. This presentation is an overview of the sands included and a short discussion of the issues involved in such a project.

A selection of four sand images from the *Playing in the Sands* Poster are provided below:

Bretignolles-Sur-Mer, Vendee, France. Sapphires in a heavy mineral sand plus garnets, zircon, quartz, all well polished	Barren River Lake, Edmonson, Kentucky. Marine Fossils, bryozoans, crinoids, corals. Ordovician – Devonian 300-450 Million years old

Swimming Hole, Sinapalo, Rota Northern Mariana Islands.
Star forams and red homotrema (foram)

Yellowstone River, Terry, Montana.
Sedimentary rocks "baked" /metamorphosed by fire in nearby coal seams

Conclusion. We enjoy producing educational material for everyone. We welcome your suggestions, collaboration and the opportunity to photograph and document sands that you might have available.

About the Speakers

Leo Kenney has degrees from Boston College and Harvard University, yet he received his real education teaching high school biology for 35 years. During this period his scientific macro-photography was published in numerous texts and publications including *Field Guide to Animals of Vernal Pools* of which he is a co-author. His sand photography has been featured in several magazines and textbooks and has been the staple of the Sand Calendars for 10 years. His slide show skills have been honed by feedback from thousands of critical teenagers.

Kate Clover has been a geologist since about age 4, picking up pretty rocks, agates, and fossils along the beaches. She grew up on a pure white "sugar sand" beach on Lake Michigan and began to wonder about how sands are different when she traveled to a dark rocky beach on Lake Superior probably before she turned 10. Rocks, fossils, plus picking copper and iron samples from poor rock mine piles in Upper Michigan led her to Michigan Technological University to study geology. Retired now, she continues to be amazed at the beauty of sands when viewed under the microscope and all the geology and marine science that sands can teach.

Sand Toys: Beyond Pails and Shovels

Lori Carter, Member ISCS
Sigma Software, Inc.
Roswell, GA

Introduction
The topic of sand toys may invoke images of little sand pails and shovels used to build sand castles, but there is considerably more to the involvement of sand with toys. My experience with sand toys is primarily from what can be found on the internet, in books, and in antique shops, so the history as I present it here is far from complete, but fascinating nonetheless. All of the toys in this article are from my personal collection.

Ancient Toys
Marbles have been reported as far back as 3000 BC[1] and sand may have been used as a playing board for those games. Early doll bodies and balls were often filled with sand. These were rather simplistic uses of sand and there does not seem to be much more mention of sand in the history of toys for a very long time.

Sand Automata: Early to Mid-1800's through 1970's
Suddenly, in the early to mid-1800's, a great leap in the sophistication of sand toys occurred. Automota, i.e. mechanical figures, had been around for centuries, and even sand-powered automata were documented as early as the 1st century AD, but industrialization in the Victorian era brought these curious devices to the masses [2]. People in England, America, France, and Germany were entertained by sand-powered automata featuring acrobats, dancers, musicians, and other moving figures. Elaborate sand automata made of wood depicted whole scenes and operated multiple figures. Movement was driven by sand falling onto a paddle wheel that turned an axle attached to a jointed figure.

Though they enchanted young and old alike, Victorian sand automata were probably more of a parlor curiosity and not really meant for tiny hands. As the Victorian era ended, their popularity waned, but a few notable sand automata were produced commercially during the 1900's.

Figure 1. This sand automaton (maker and date unknown, possibly early 1900's) is constructed of cardboard rather than wood, but still employs the same sand mechanism as its Victorian counterparts. It depicts a strongman displaying his gymnastic strength and skill. Scenes around the toy appear to be military and include tents, men on horses, and a baseball game.

There is a paucity of sand automata until the 1950's, when a rare sand automaton called "Nutto" appeared. It was made of cardboard shaped like a cage one would see on an old circus train. In it, a monkey made of rubber performs fancy flips behind metal bars for an admiring audience of fellow simians. Like Victorian sand automata, Nutto is powered by a sand wheel mechanism.

Then, in the late 1960's, a spate of modern reproductions of Victorian sand automata emerged. Like "Nutto", they were made of cardboard instead of wood. There was even a plastic version licensed by the Walt Disney Company in the late 1970's. It is the most recent mass-produced sand automaton I have found as of this writing.

Figure 2. The Shackman toy company was probably the most prolific producer of modern sand automata (circa late 1960's), depicting either high-flying acrobats or a grinning organ grinder cat that is inspiring a dancing mouse to twirl gleefully around.

Figure 3. A sand automaton by Vandor Imports (1968) touts "Alfred the Amazing", "the Flying Marvel", an "Acrobatic Bafflement". The toy is styled as a "Dingaling Brothers" circus tent, complete with ringmaster on the back drawing a crowd with his "Laaydees and Gennelmun" side show barking.

Figure 4. A Disney sand automaton (circa late 1970's) features Donald Duck and Mickey Mouse as the sand-powered acrobats. The back of the toy is clear, so you can see how the sand operates the toy.

Now, like their Victorian predecessors, contemporary sand automata are not considered toys. They exist mainly in the world of art and are no longer produced commercially. Atlanta artist Betsy Browning has revived sand automata by creating beautiful, fanciful renditions that can be seen on YouTube. Author Walter Ruffler, in his 2011 book, "Paper Models that Move", provides detailed instructions for building various automata from cardstock, including sand operated models. He also has a website where one can purchase kits to make sand operated automata and other toys. It is encouraging to see these sand toys have not been forgotten.

Sand-Powered Toys: Early 1900's to Present

From the early 1900's through the early 1950's, sand toys made for children became popular in the U.S. Some were made of wood or cast iron or even cardboard, but at that time, toy manufacturers were experiencing a boom in demand for tin lithographed toys. Most of the toys had a funnel or hopper into which one would pour sand. Then, various actions would occur.

Figure 5. Some sand-powered toys were made of paper, cardboard, and wood. This elephant toy (maker unknown and age unknown, but possibly early 1900's) has a sand hopper that drizzles sand into a scoop on the elephant's trunk. The weight of the sand eventually tips the elephant's head down until it dumps the sand. The head is counter-weighted so it returns automatically to an upright position to receive more sand.

Sand-Duz-It Toys: 1910's, 1920's

Not much information is available about San-Duz-It sand toys other than a 1920 advertisement and a brief mention in the January, 1917, Volume 46 issue of "The Bookseller, Newsdealer & Stationer" periodical in an article about the 1917 toy fair. It says the toys were made by the Gopher Toy and Novelty Company in Minneapolis, Minnesota.

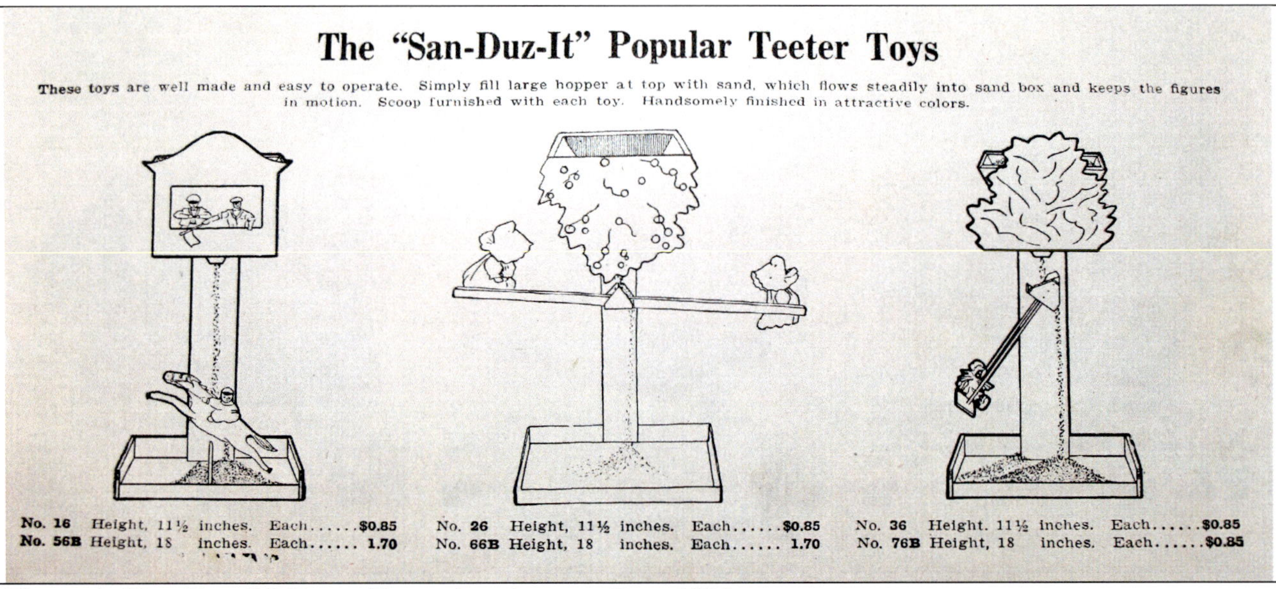

Figure 6. "San-Duz-It" Teeter Toys advertisement from 1920.

Wolverine Toy Company: Early 1900's to 1950's

The Wolverine toy company was the most prolific sand-powered toy maker. They produced numerous "Sandy Andy" toys after acquiring the line from the Sand Toy Company in 1909. Some of their toys were simple devices powered by sand and automated by counterweights. Other toys used similar mechanics but were more sophisticated and often comical or whimsical. (Wolverine history and toy dates were compiled from the book "Wolverine Mechanical Toys" [3])

Figure 7. "Sandy Andy Incline No. 60" (circa 1917-1932)

Figure 8. "Sandy Andy See-Saw No. 61" (circa 1927)

Figure 9. "Sandy Andy Incline No. 76a" with engineer (circa 1927-1932)

Figure 10. Pirate themed "Captain Sandy Andy" (circa 1936)

Most Sandy Andy's used gravity and the weight of the sand to move a cart or character until the sand spilled or was dumped, then a counter weight would reset the toy for another round.

Figure 11. The Wolverine "Merry Miller" (circa 1929-1932) features the eponymous miller holding a big key and followed by his dog. As he creeps around the side of the building, his rolling-pin wielding wife waits around the corner in a particularly perturbed state. Right before the miller reaches the edge of the building, he suddenly snaps back, just in time to avoid a knot on the head and ready to begin creeping forward again.

Figure 12. During WWII, the Wolverine company contributed its resources to the war effort. Yet, because of the efficiency of new machines, they were able to continue producing toys with non-critical materials instead of tin. This sand mill is made of reinforced paper, cardboard, and wood. It depicts Dutch children pulling each other back and forth, powered by a paddle wheel mechanism. Perhaps this toy was not popular, because this design does not appear to have been continued as a tin lithographed toy after the war.

Figure 13. The Wolverine "Merry Masons" (circa 1951), features construction workers on the side of a brick building passing sand from one worker to another. In the close-up on the right, one of the workers has his foot stuck in a bucket.

Figure 14. The Wolverine Sand Loader (circa 1951-1953) drizzles sand from a hopper into a bucket on a loading arm. When the weight of the sand pulls the bucket away from the hopper, a trap door closes to stop the flow of sand. The arm swings out until a string stops it and pulls opens a door on the bucket so the sand is dropped. A counterweight pulls the arm back to receive another scoop of sand, and the process continues until the sand hopper is empty.

J. Chein & Company: 1930's through 1950's

Though more well-known for other tin lithographed toys, J. Chein & Company produced sand toys as well, including a line called "Busy Mike". These toys often featured children or animals. There were "see-saw" toys with children or rabbits; a man or little boy holding up a sand hod to catch then dump sand; a monkey turning a crank; sand chutes; and various sand mills where sand would turn one or two wheels.

Figure 15. J. Chein "Busy Mike" monkey (circa 1930's or 1950's) When sand pours through the funnel, the monkey jumps into action turning the wheel. Or, maybe the wheel is turning the monkey…

Figure 16. J.Chein Sand Mill (circa 1920's). This toy has a simple design. When sand flows through the funnel, it simply turns the wheel.

Figure 17. J. Chein Sand Chute (circa 1930's). This toy has a more sophisticated design than other J. Chein sand toys. As sand flows through the funnel, it drops into the cart until the cart is full. As the cart travels down a rail, it pulls a cover that stops the flow of sand. When the cart reaches the bottom of the rail, it tips forward so the sand flows out. Without the weight of the sand, a spring on the other end of the rail pulls the rail back up so the cart returns to its starting point causing the cart to push the funnel cover open to restart the flow of sand from the funnel into the cart again.

Figure 18. Early J. Chein see-saw toy featuring children on hobby horses (circa 1920's)

Figure 19. Later J. Chein see-saw sand toy (circa 1930's)

Rowley Toy Company: 1950's

The Rowley toy company produced "Rhythm Educational Sand Toys" in the 1950's. The "Village Blacksmith" pounded away on an anvil, a cowboy on a horse made a "Klippity Klop" sound, and a clock made a "Tick-Tock" sound. All three toys had a fluted funnel that drizzled sand onto the mechanism that operated the toy.

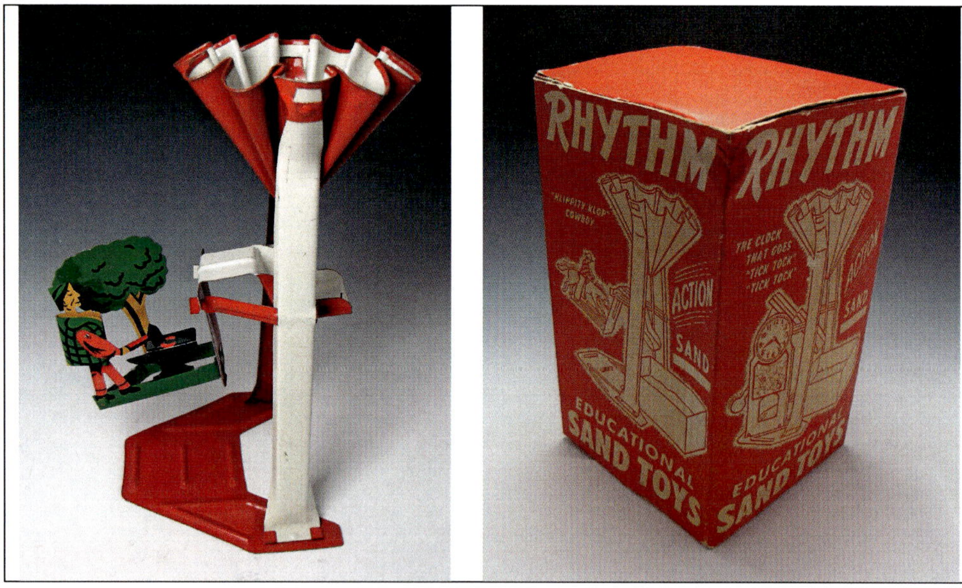

Figure 20. As sand pours through the funnel, the Rowley "Village Blacksmith" bounces up and down causing his hammer to pound the anvil. Each Rhythm toy was packaged in a box that advertised all three Rowley toys.

Windmill Toys: Early 1900's to 1960's

Windmills were popular in sand-powered toys. Most contained a paddle wheel that rotated as sand fell onto it and subsequently turned the windmill vanes.

Figure 21. This windmill has scenes depicting children playing with the windmill toy itself. Maker and age unknown.

Figure 22. "Mac Dutch Mill" made by the McDowell Toy Company (circa early 1900's).

Figure 23. This windmill made by the T. Cohn company (circa 1940's or 50's) can be folded flat for storage.

Figure 24. Plastic "Sand and Water Mill" (1968), Sydney A. Tarrson Company.

By the 1960's, sand-operated tin lithographed toys were no longer being produced as other mechanical tin toys edged them out. Starting in the late 1950's through the early 1960's, the tin toy market itself gave way to the plastic toy market. Plastic toys were easier to make, inexpensive, and did not have the sharp edges that tin toys could have. Though plastic sand-powered toys continued to be made from the 1960's forward, the heyday of sand-powered toys was essentially over.

Other Sand-Powered Toys: 1980's and 1990's

The "Sand Flipper", made in the early 1980's, is reminiscent of the Wolverine "Merry Masons", but without any characters. It simply passes sand from bin to bin, but a Disney version made around the early 1990's inexplicably has princesses passing sand down a castle wall.

Figure 25. Sand Flipper (circa early 1980's)

Figure 26. Disney Princesses Sand Toy (circa early 1990's)

Other Sand-Powered Toys: Present

Today's sand-powered toys are barely more than a funnel with a wheel or two to turn. Though they are cute and have bright colors, they aren't very sophisticated. But, there are a few exceptions that feature multiple moving parts with engaging characters, so maybe there is still some hope that this genre of sand toys will maintain a glimmer of its glory days.

Figure 27. The Disney "Pooh 100 Acre Wood Sand Sorter" (2007) has a double funnel. One side sends sand pouring through Pooh's honey pot onto a wheel. The other side turns a wheel that rains sand onto Tigger's umbrella.

Figure 28. "Owl About Waterfalls" (2014) has an owl cup for scooping sand into the top funnel, a tilting mid-toy funnel, three wheels for falling sand to twirl, plus a little dump truck for catching sand.

Other Sand Toys: 1950's to Present

In 1955, the Smethport Specialty Company discovered a use for dust created during magnet production. The magnet grinding dust was incorporated into its iconic "Wooly Willy" magnetic drawing toy (1955 to present) [4]. Smethport also produced the "Magnetic Sandbox" (1967) and "Magnetic Dozer" (circa 1973), similar toys that use a magnet to control a bulldozer that pushes "sand" around. The sandbox toy has fine plastic crumbs or shavings as sand, and the dozer toy has small brown plastic pellets as dirt.

Figure 29. "Wooly Willy" (1955 to present) uses sand-sized magnet dust.

Figure 30. The "Magnetic Sandbox" (1967) has a bulldozer for pushing plastic "sand".

1960's

A curious set of Cracker Jack prizes appeared around 1960. Tiny cards with two plastic windows incorporated into a small picture also contained sand. The sand flows from one window to the other, then when the card is turned, the sand flows back. A series of 14 cards included an odd assortment of subjects, e.g. a kitten getting a belly full of milk from a pitcher, a dump truck, a soda fountain, a water pump, a sand castle, and a jet being refueled mid-air by another jet.

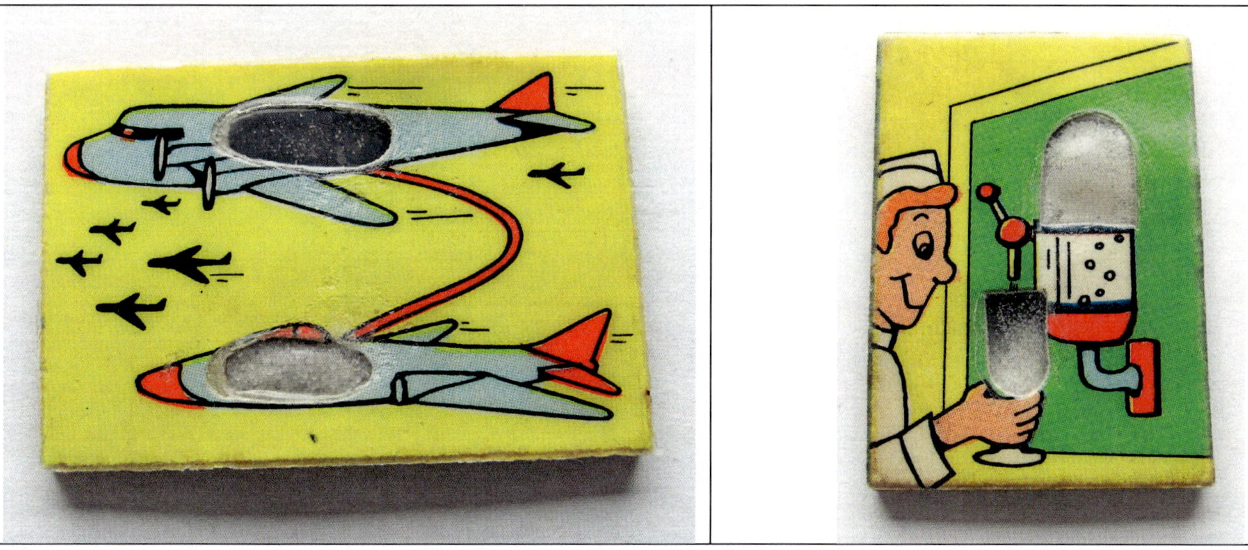

Figure 31. Cracker Jack sand toys like these (1960) are rare because they were relatively delicate and imminently destroyed by small hands.

1970's

The Funtastic company made a product called "Silly Sand" in the late 1960's, early 1970's. First, you mix sand with water in a plastic squeeze bottle. Then, you squeeze the mixture onto a special mat to build mountains, sand castles, and other crude structures. When the structures dry, you can knock them down and use the sand again. From 1973 to 1979, the Wham-O company offered the "Magic Window" toy. Unlike similar desk toys, Magic Window does not contain water. Though marketing for the toy claims it contains "microdium" crystals, there is little information as to the exact composition of the particles inside the toy.

Figure 32. "Magic Window" (1973 to 1979). The patent indicates the "microdium" crystals are roughly sand-sized particles, each color having particular physical characteristics relative to size, shape, and/or density. These differences prevent the colors from becoming a homogenous mixture. The crystals move when the window is tilted, and because the colors do not mix together, interesting swirls and other patterns can be formed. [5].

1980's and 1990's

In the 1980's, hydrophobic sand was marketed as "Magic Sand" by Wham-O. Incredible underwater structures could be built with Magic Sand, then the sand could be removed from the water, completely dry, ready to be used again.

In addition to toys, there are a couple of games where sand is integral. "Quicksand: The Sand Race Against Time", was produced by the Whitman company in the early 1980's. It uses sand timers as game pieces on an hourglass shaped board. In the 1990's, the TOMY toy company, perhaps best known for its "Furby" electronic toy, made a hand-held game of skill called "SandSationals".

Figure 33. "Quicksand: The Sand Race Against Time" (circa 1980's). Hourglass game pieces had to be moved from one side to the other before the sand ran out.

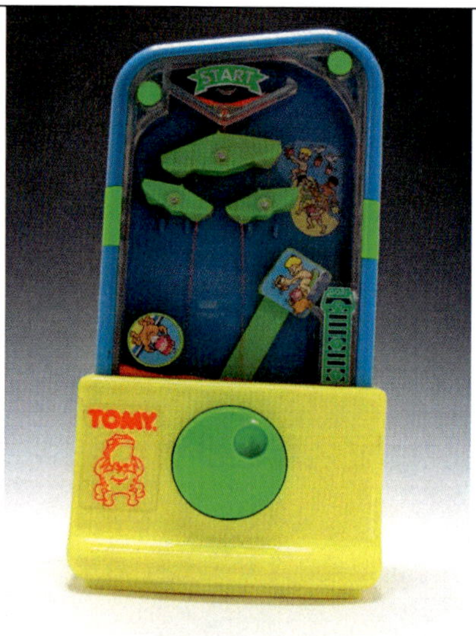

Figure 34. "SandSationals" (circa 1990's) challenges the player to fill a vessel at the bottom by capturing sand spilling from the top onto tipping platforms.

2000's

Cracker Jack used to have the occasional plastic hourglass, and gumball machines may still have some even now, but some surprising sand hourglasses delighted children in Europe around 2009. Kinder Surprise (also Kinder Egg), made by the Ferrero candy company, are hollow chocolate eggs with plastic toys inside. Their series of hourglass toy prizes has five different colors of sand, and each color has a different animal printed on the outside.

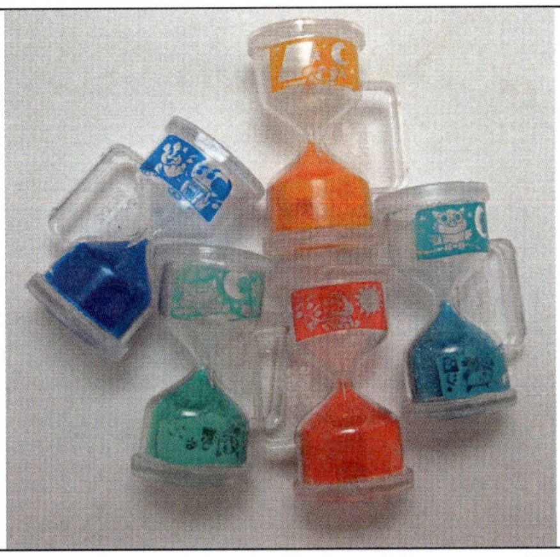

Figure 35. Kinder Eggs hourglass toys (2009).

In 2013, several companies started marketing products with approximately 98% sand and 2% polymer similar to "Silly Putty" [6]. With names like "Kinetic Sand", "Moon Sand", "Sands Alive", and "Squishy Sands", the sand can hold a shape briefly after it has been squeezed or molded. Also popular for its sensory attributes, this sand toy continues to be popular at the time of this writing.

Figure 36. Sand mixed with a polymer (2013) stays together and won't spill like sand alone.

Other sensory "stress" or "fidget" toys feature sand, though the sand tends to be more incidental rather than the power source or main focus of the toy. Sandboxes, educational sand tables where children can experiment with sand using various objects, computer games that feature digital sand, and blocks with sand inside are other examples of modern sand toys. We have even come full circle with sand "stuffed" animals and certain "hacky sack" foot bag balls filled with sand.

And lest one question if sand is actually used in toys, there are some that allow a peek. Wooly Willy's magnet dust is angular but appears to be more uniform and rounded than one would expect. The Magnetic Sandbox clearly uses plastic for its "sand" and the Sandsationals game has tiny glass spheres. The Quicksand board game pieces appear to have very fine sand and acrobat automatons definitely use real sand. The Disney automaton even has biogenic sand, perhaps an indication of where it was manufactured – Hong Kong!

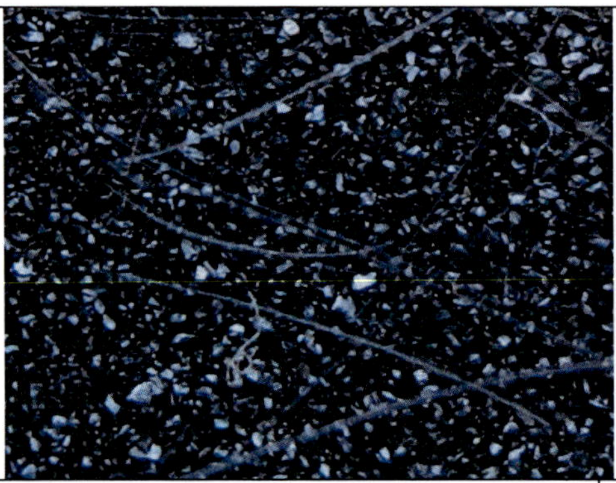		
Figure 37. Wooly Willy magnet "sand" (Figure 29)	Figure 38. Magnetic Sandbox plastic "sand" (Figure 30)	Figure 39. Sandsationals glass sphere "sand" (Figure 34)

Figure 40. Acrobat automaton sand, lower right (Figure 2).	Figure 41. Acrobat automaton sand (Figure 1)	Figure 42. Disney automaton biogenic sand (Figure 4)

Conclusion

Sand has played an important role in toys and continues to make an occasional appearance, though the use of sand in toys has declined over time. The engineering sophistication of sand toys has also declined. Perhaps because the children of today tend to be more engaged by electronic devices or because toy makers have relegated sand-powered toys to the tiny tot market, the sand toys of antiquity are not likely to see a resurgence. But should you encounter one, take some time to examine it, see how simple yet ingenious it is. And play with it.

References

1. Webpage: "The Museum of American Glass in West Virginia: The Origin of the Game of Marbles" by Chris Cooper; URL: http://www.magwv.com/magmarble/origin_of_the_game_of_marbles.html (accessed April 13, 2018)
2. Webpage: "World Encyclopedia of Puppetry Arts"; URL: https://wepa.unima.org/en/automata-androids-and-robots/ (accessed April 4, 2018)
3. Terry Mowrey (2013) *Wolverine Mechanical Toys: A Collector's Journal and Numbered Catalog List*, Word Association Publishers, PA, 186 pp.
4. Webpage: "Smethport History"; URL: http://www.smethporthistory.org/smethportspeciality/wollypage.htm (accessed April 14, 2017)
5. Webpage: "History of the original Wham O Magic Window"; URL: www.magicwindows.org/history.htm (accessed October 10, 2017)
6. Webpage: "International Science Times"; URL: http://www.isciencetimes.com/articles/6100/20130920/silly-putty-sand-kinetic-new-toy-stress.htm (accessed April 14, 2018)

About the Speaker

Lori Carter has been collecting sand for many years, but she has been collecting sand toys for only a few years. Her interest is mainly in vintage or antique sand toys, though she has several modern sand toys in her collection because some day they will be old too. She works hard to uncover the history of the toys in her collection and is a pure amateur in those endeavors. Lori has a BS in Computer Science from the University of West Florida and continues to work as a consultant in Information Technology. She has been a long time member of the Georgia Mineral Society and volunteers in many capacities for the club, speaking about sand at various venues and becoming the Sand Princess to teach about geology through sand at elementary school STEM nights. She is also a member of the International Sand Collectors Society and has helped with the ISCS SandFest conferences over the years. Her publishing credits include several articles in *Rock & Gem* magazine.

Sand Dune Garden Shadow Boxes

Frances L. Peters
Member, International Sand Collectors Society
Eastern Shore, Maryland

Updating My Sandbox. My interest in sand collecting got serious after a friend from my "heart-shaped rock" collecting days sent me the most unusual sands. One sample was full of little stars. I started asking for gifts of sands from family and friends.

Figure 1. My Sand Box. It currently contains samples from 12 states and 4 countries.

My collection grew and grew as friends and acquaintances brought samples back from their vacations. Figure 2 illustrates some of the colors and textures that were gathered.

Figure 2. These field samples are easy to handle and study. (Each diamond = approx. 2"x 3")

Locally, my friends from work provided sand samples from the rivers of the Delmarva. I also took many day trips to hunt for and collect sand from the shores of the Chesapeake Bay. Figure 3 shows the Delmarva map and samples. These samples will be studied using the microscope shown in Figure 5 with the hopes of finding a few heart-shaped grains of sand.

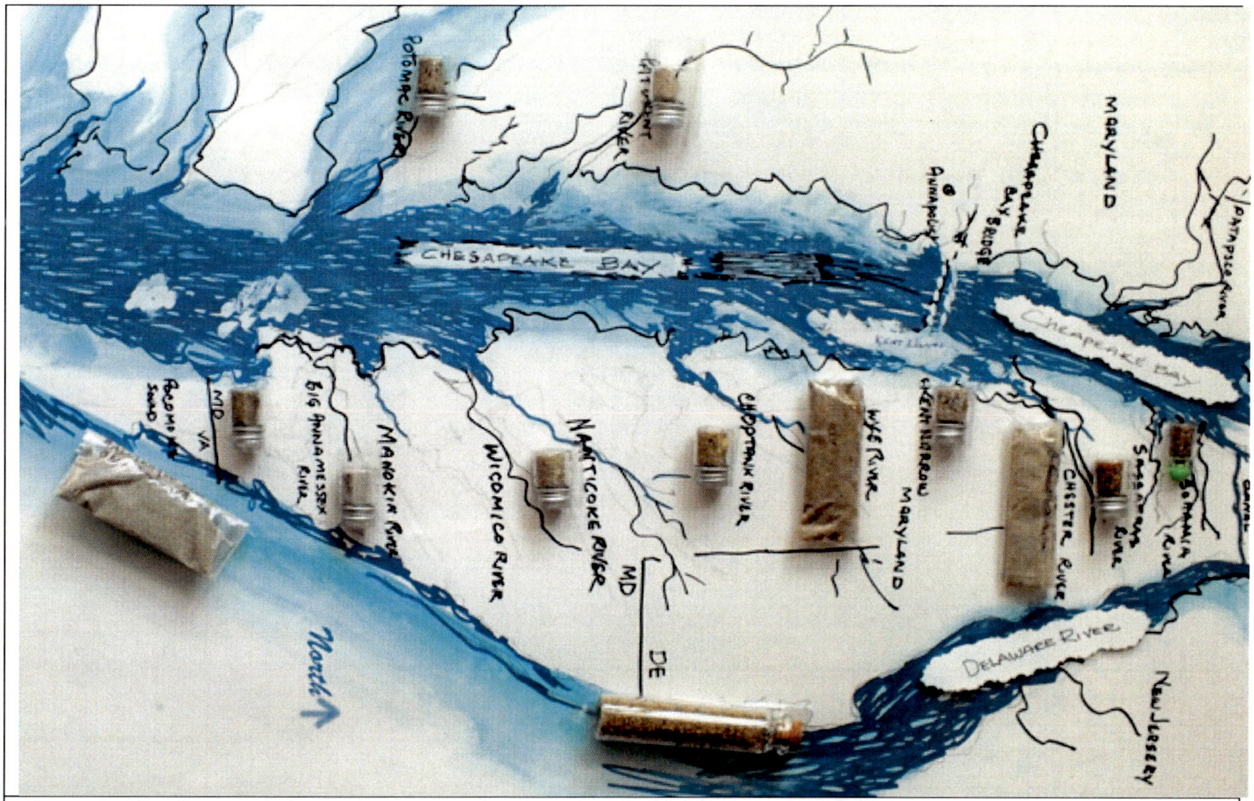

Figure 3. Simply stated, the Delmarva Peninsula is bordered on the east by the Delaware Bay and on the west by the Chesapeake Bay. It is all in the Atlantic Coastal Plain. Some of samples are shown at their collecting sites.

My first sand dune garden boxes were crafted for a sea glass festival held at one of our local Chesapeake Bay environmental centers. The idea of the box was to have a place where you could display the best things you found on the beach.

Figure 4. This is an early Sand Dune Garden Shadow Box. It is mixed media using an original photo for the background with applied sand, a found object and a handmade wire-wrapped tree.

The Process. My artistic process is about exploration, discovery, investigation, imagination, experimentation and expressing delight and wonder at what is found.

My workshop. Being able to see the sand enlarged is so much fun. I can still remember the thrill the day I discovered my first heart-shaped grain of sand.

Figure 5. I place the grains of sand on black paper, put the lighted microscope on top of the sand, and look to see what's there.

My filing system. Consists of multiple three ring binders. The enlarged samples are printed then placed in clear sheet protectors along with the sand sample cards and filed in one of the binders.

Printed enlarged sand from sample card in binder	Sand samples shown with containers

Figure 6. Pages from the sand file binders showing some of colors and a heart-shaped sand grain (indicated by the dashed oval at the upper right).

My art work. Like life, my art work continues to change. It is a good thing that artists are taught to know the rules and when to break them. For it was to the discovery of sandpaper and the full range of colors it had to offer my palette, that I could not say no.

Figure 7. The beauty of sandpaper colors.

My first sandpaper sand dune garden box is pictured in Figure 8.

Figure 8. As a former wood carver, the natural patterns of the wood grain remains a source of inspiration. It was the grain that set the mood for this creation about an eagle soaring along the dunes as the evening sun slowly disappears in the western sky. The little wire-wrap tree is a symbol of hope.

Experiments with wire-wrapped sea glass, photography, and different methods to apply sand to various surfaces took place over the next years as my Sand Dune Garden Shadow boxes evolved.

The palette moved to natural sand colors only. In Figure 9, the distant dune is created with sand from the Chesapeake Bay at Matapeake beach. The darker middle ground sand is from the Delaware River. The foreground white sands are from Florida. The beach pebbles are from my collection.

Figure 9. The tree of hope. After a long winter's rest, the tree springs back to life.

It is hard to explain, but the all-natural colors joined with the craft store colored sand to express the feeling of a day on the beach in Figure 10.

Figure 10. Can you hear the sounds of the water lapping the shore and feel the wind on you face? At times my work becomes a bit abstract, like looking thru the lens of the camera up close and personal at a small section of nature.

Conclusions. So, now you are up to date on my sand box. I want to thank all my friends with the International Sand Collectors Society and *Rock & Gem* magazine for the wonderful influences you have had on my art and my life.

About the Speaker

Figure 11. Curiosity keeps me wondering, what else is out there?

As I enter my ninth decade, waking up in the morning to explore another day is the accomplishment. Currently, I work part time at our local visitor's center and volunteer at the monthly historic society open house. It is enjoyable to meet other explorers.

Figure 12. Maybe I'll make a sand castle today.

"Sand collecting is a great hobby." It keeps me active and young at heart.

Display Your Sands from Around the World

Bill Beiriger
ISCS Member
Livermore, California USA

Introduction. The purpose of this presentation is to show how simple it is to make a display with your sand samples so that others may become interested in our hobby. Following the 'how-to' portion of this presentation, ideas will be provided for developing a theme-based sand display. A computer was used to prepare most of the labels and layouts. The following topics include:

- Making liners for a display case
- Ideas for risers to put samples on.
- A simple way to make labels
- Containers for sand samples

Making a Display. First, decide what type of case you want to use. These include:

- An upright case
- A slant front case
- A flat top case
- A glass museum's show case for gem and mineral shows

When showing your collection at a gem and mineral show that allows guest exhibitors, make sure you check on the allowable display case dimensions. Sign-up can be several months in advance. The host club may have display cases you can use if you bring your own liners and risers.

Display cases and liners. Figure 1 (a) and (b) shows upright cases. They show two different ways to arrange your risers and specimens. Figure 2 shows two slant-front cases, both having no risers. Both show labeled specimens without risers, but in (b), there is also some additional information on the rear wall.

One way to make a fabric bottom liner for an upright display is to use a sheet of door skin wood or cardboard. Place the cardboard on a piece of fabric that is lying face-down on a flat surface, leaving a border sticking out around the edges. Fold the opposite ends of the fabric over and tape them to the wood or cardboard. Fold the other two sides over and tape them down as well. Repeat this procedure for the back and end panels of the case. A similar procedure is used to make fabric liners for a slant-front case, but in this case, the shape of the ends should include the slant. Again, place the door skin wood or cardboard over the fabric that is cut larger to allow it to stick out all around. Fold and tape. Repeat for the panel at the opposite end, the back, and the bottom.

Figure 1. Upright cases with different styles of risers (photographs by the author from mineral shows).

When preparing fabric liners, the color is optional but do not overpower the samples with bright colors. Also, black and white liners are not recommended. Black shows dirt and dust and gives a lot of reflections in the glass. White will also show a lot of dust and dirt. Try neutral colored fabrics: gray, light blue, light green, or most earth tones. Draped fabric liners are fast to make but their wrinkles and creases can be very distracting.

Making risers. Making risers is similar to the way liners are made. The only difference is that the liners are 1, 1-1/2, or 2 inches high. You can make boxes out of wood or foam board. Like the liners, wrap the fabric around the back and tape it. One approach is to make a stepped stack of three or four riser boxes of equal height. For the three-step version, the bottom one is 24 x 48", the middle one is 18 x 48", and the top riser layer is 9 x 48". Options also include a rectangular bottom riser with triangular ones at the rear corners. If sand is in different sized containers, the taller samples should be placed at the rear and shorter ones in front.

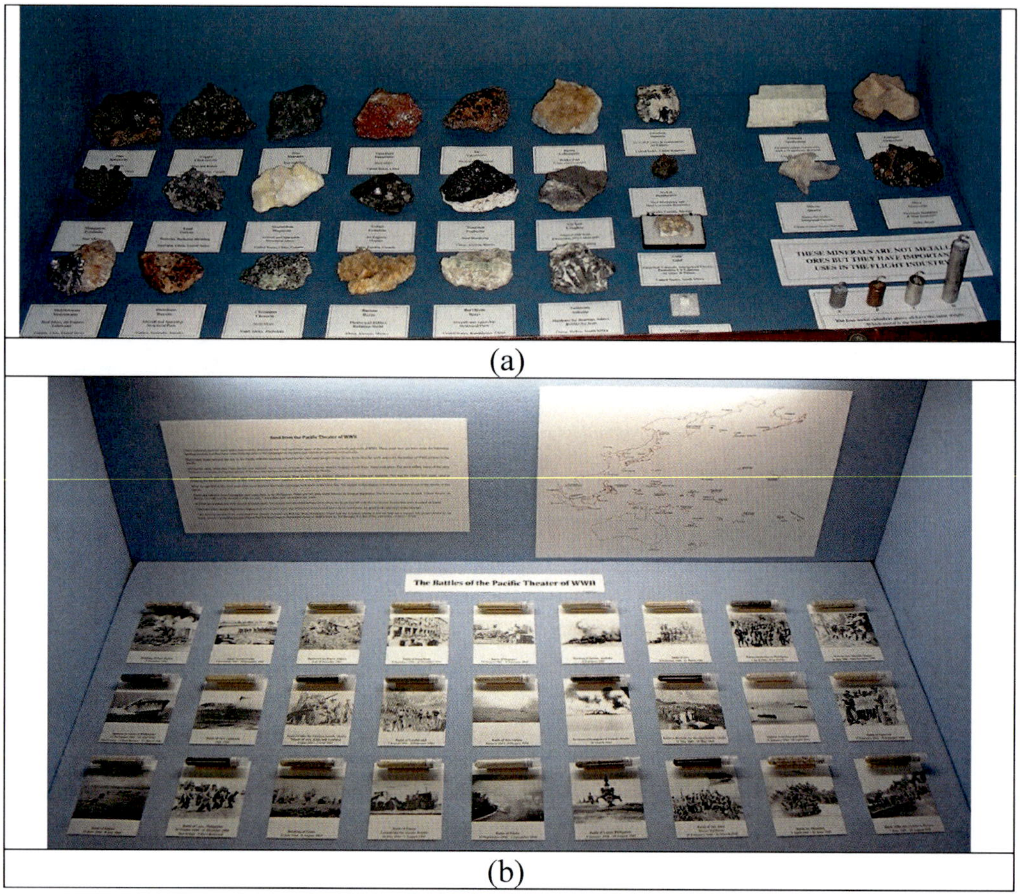

(a)

(b)

Figure 2. Slant front cases without risers.

Labels are the most important part of any display. Like the case, there are a variety of options for labels. A simple one can contain only the sample's collection location, but others can include a photo of the location. Other label details can be: what the sand is made of, the geological source of the samples, why the grains are shaped as they are. You can also include photomicrographs of the samples to help the viewer to learn something from your display.

Label software. I makes all my labels and layouts using Adobe Illustrator™ (AI) software but it has some drawbacks the reader should be aware of. There are similar programs with layout capabilities available from other companies. A major problem with the current AI is that, at this writing, it is part of a subscription service. The software can be used only as long as the subscription remains valid and access to your files will be lost if the subscription becomes inactive. The cost of use can be very pricey.

The 'layers' part of AI is what I use for display labels. Each of the following parts that make up a label consists of its own layer:

- Layout lines
- Outline
- Photograph outline
- Photo insert
- Text
- Ability to make the layout visible or invisible

- A locking function to keep changes from being made to a given layer while you work on another layer or layers

Several step-by-step examples are presented in this talk to explain how AI in the 'layers mode' can be used to create specimen labels. Figures 3 (a) and (b) show a simple label with rounded corners and a more complicated multi-layered label containing a photo of the collecting location.

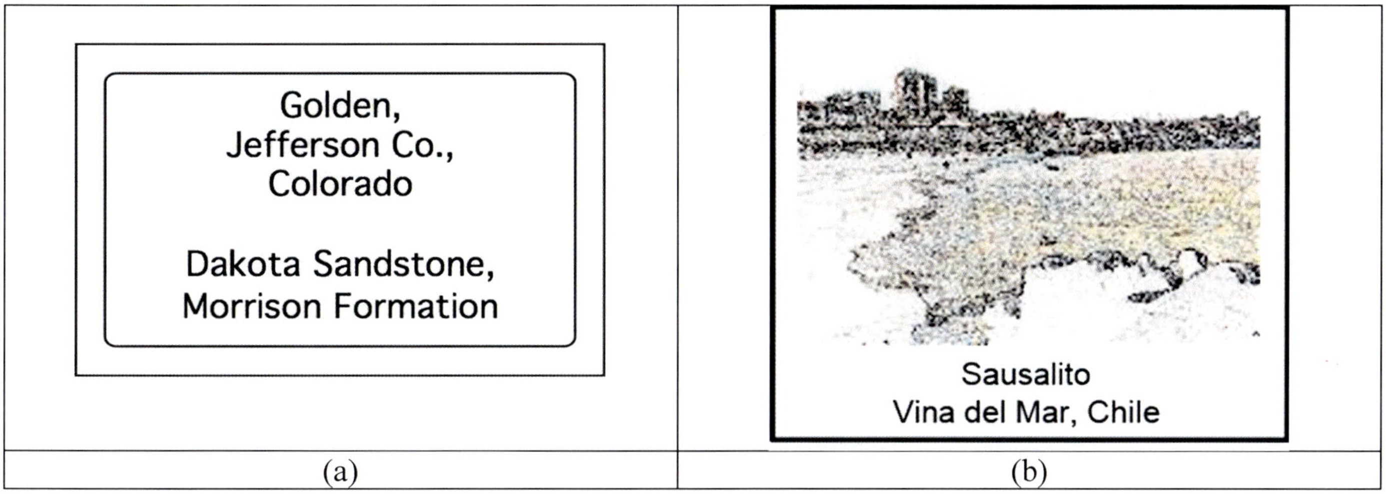

Figure 3. Examples of a simple specimen label (a) and a photo label (b).

Setting up the display. After preparing the case, risers, and labels, it is time to set up the display. First place the labels into position on the risers and then place the sand samples. Most displays should have a minimum of 20 samples. Step back to inspect the display to make sure everything is in the right place, that the labels are lined up, and that any dust or foreign matter has been removed. Examples of a number of arrangements including the following are discussed in more detail in the author's talk:

- A liner with 3 or 4 risers with rows of either round or square sand containers
- A display with no risers but larger photo-type labels and wider spacing of square contains
- Fewer specimen containers but with more educational content (maps, history) about each one on the back panel.

Layouts can be prepared by using guide lines that are scaled to the length, width, and dimensions of your design for each riser or platform. Spacing can be square or off-set. Figure 4 shows a layout with guidelines superimposed (a) and without the guidelines (b). This tests how your display could look and you can rearrange it as you wish.

Double-sided tape can be used to save time and keep labels in place. Do not mark the liner, but you can use a meter or yard stick to ensure the proper distance between labels. If possible, vary the colors and textures of samples.

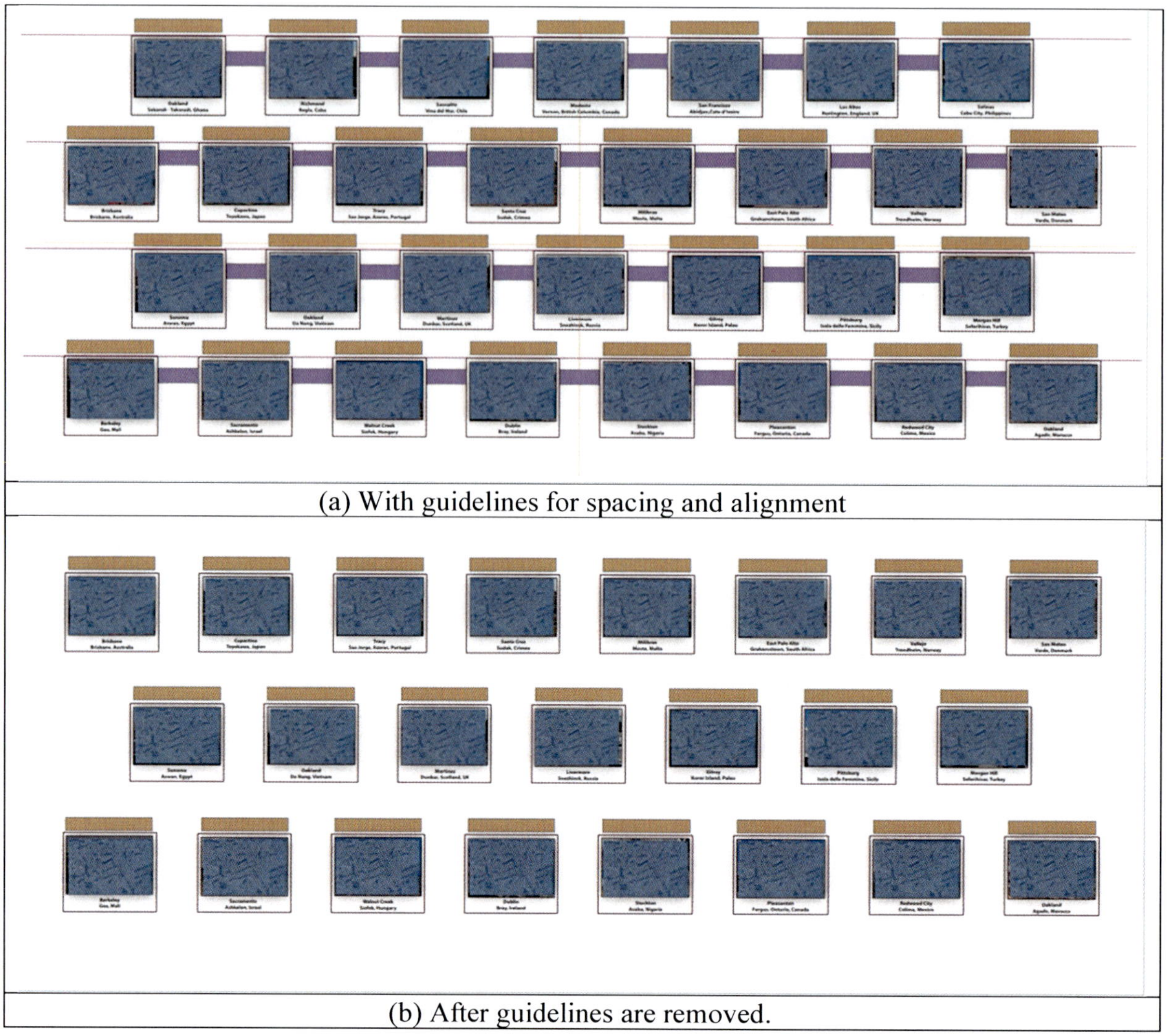
(a) With guidelines for spacing and alignment

(b) After guidelines are removed.

Figure 4. Arrangement of picture labels with and without guidelines.

About containers. A wide variety of shapes and sizes of sample containers is available. Some are available from mineral and gem suppliers (plastic boxes with lids), others from science equipment suppliers (glass vials), and others from hobby and decorating stores. Note that jars with heavy glass walls may distort the view of the sand from some angles.

Themed presentations. Several themed displays were shown in the presentation. They include, for example, a 'non-educational' fun collection of sand with vacation souvenirs. The author's sand collecting experience started with "Micro Fossil Foraminifera" and the next was "Very Colorful Sands". The first major competitive sand display was "The Terminology Used to Describe Sand". It was presented at the California Federation of Mineralogical Societies (CFMS) and the American Federation of Mineralogical Societies (AFMS). There are seven regional federations in addition to the AFMS.

Competition judges will find your mistakes, so when setting up watch for these things:

- Errors in label spelling, geological formation identification
- Alignment of labels or mislabeled specimens

- Liners with wrinkles or creases
- Gaps or misaligned panels or risers
- Dust and foreign matter

And remember, educational displays should contain easy-to-understand information.

Some possible display themes are: local area history, U.S. cities from the Civil War or Revolutionary War, sister cities for towns in your area, cities for little league teams from the U.S. and around the world, locations from your favorite books. The author's themes have also included these:

- TV and Movie Filming Sites
- Troop Movements and Beaches in Sicily and Italy During WWII
- The Pacific Theater of WWII from the Aleutian Islands to Australia
- U.S. Route 66 the "Mother Road"
- The Lewis and Clark "Trail of Discovery"
- Sister Cities for Towns in Northern California

Conclusions. This extended abstract has illustrated how to make displays, labels, and risers for your sand collection and suggests possible themes. ISCS members are encouraged to produce great sand displays. Be sure to include your contact information along with your displays to find new trading partners or to attract new ISCS members. Tip: When giving e-mail information, make sure the sender includes "Sand" in the subject line so you know it is not junk mail.

About the Speaker

Bill Beiriger worked at Lawrence Livermore National Laboratory for 27 years. During most of his time there he did analytical X-Ray Diffraction, Electron Microprobe, Scanning Microscopy and Thin-Sections preparation for Optical Microscopy.

He has been a member of several gem and mineral societies for more then 50 years. During that time he has competed with educational displays and displays of micro-mount minerals. He has won trophies in the California, Northwest Federations and the American Federation of Mineral Societies.

Bill started collecting sand samples in 1955. In 1999 he corresponded with Rex Elliott, a member of ISCS in 1999. Since then his sand collection has increased from 110 samples to almost 4700.

Scottish Sand Stories

Michael R Hughes BSc, PgDip, Member ISCS
Retired Nature Reserves Manager
Cupar, Fife, Scotland

Introduction. This video is based on a presentation made to my local natural history society. Suggesting a talk on sand had elicited the usual range of reactions, from polite disinterest to downright mocking, but the mood changed after I'd handed round a couple of prints of images of sand through the microscope. Nonetheless, the challenge was still there to make the talk interesting to a lay audience so I chose sand which had stories to tell.

Scotland is a good place for a story-telling arenophile. For a small country it has a wide range of fascinating geology and coastal morphology, mostly within half a day's travel from the centre. Add to that a knack of producing eminent geologists, some top class Earth science departments and societies, a vibrant and well documented human culture going back thousands of years, there is an awful lot to draw on. Oh yes, and it's drop dead gorgeous to look at as well.

Why We All Like Beaches. I first realised that I possessed a groin when I was about four years old. It wasn't an enjoyable experience. My father had just coerced me into going for a deep paddle in the North Sea and the cold, wet, sandy wool of my hand-knitted bathing trunks plugged straight into that part of the brain that processes misery. Tears were not far away. But, suddenly, buckets and spades appeared and the discomfort faded, soothed away by rising excitement. I had a mission. Scarborough was going to see a sandcastle of such magnificent elegance and invention that the sea, recognising true craftsmanship, would have a word with the moon and stop coming in.

I'd be very surprised if I was alone in remembering an experience like this as the start of a life-long attachment to sandy beaches. They're the perfect place for exercise or contemplation or, hopefully on rare occasions, to grieve. Best of all though, the go-on-forever views suck away day-to-day worries and distractions leaving you with a nice clean skull, occupied only by a spirit of playful optimism and a sense of well-being from having stretched your muscles and cleared your sinuses in the cool sea-freshened air. These are the perfect conditions for creative thinking - or at least a bit of mild pondering!

The River Tay. A geology field trip to Balmerino on River Tay estuary, ostensibly to search for agates, was hi-jacked when we became more interested in the incredibly diverse range of pebbles. Rosalind, our patient and knowledgeable tutor, explained that this variety was because successive glaciers and glacial melt waters had collected, carried and dumped stone debris from a huge area covering many different rock types, some from well beyond the Highland Boundary fault.

This made me wonder what the sand was like at Tentsmuir at the mouth of the estuary. Would it be as varied? Well, I think the images below answer that question!

Pebbles at Balmerino, River Tay estuary.	*Sand at Tentsmuir at the mouth of the Tay.*

A bewildering array of minerals on view here and I thought it would interesting to see what rock types all those glaciers might have molested on their journey to the North Sea. These included (deep breath) siltstones, sandstones and conglomerates, psammites, pellites and quartzites, basalts, tuffs, a full suite of granites... [1] Tell you what, let's go somewhere else where the geology is a bit less complicated!

The Island of Eigg. Over in the west of Scotland, the bedrock geology is not quite so varied and the sands reflect this. Eigg is one of the four islands that make up the Small Isles parish in the Sea of the Hebrides. On its Atlantic coast is a broad raised beach, the crofting district of Cleadale, which in contrast to most of Eigg is well-drained fertile ground based on ancient oyster beds and sandstones. At the seaward edge, the trinity of time, chemistry and weather have created the sort of bizarre geology that attracts even non-geologists, though it was Hugh Miller, a consummate professional-amateur field scientist, who best described these curious rounded boulders:

Here and there we find them lying detached on the beach, like huge shot, compared with which the greenstone balls of Mons Meg[1] *are but marbles for children to play with ... In other cases they project from the mural front of rampart-like precipices, as if they had been showered into them by the ordnance of some besieging battery, and had stuck fast in the mason-work.* [2]

Concretions in Jurassic sandstone, Eigg.	*Traigh na Bigil, Eigg.*

[1] Mons Meg is a massive bombard that sits on the walls of Edinburgh Castle.

Miller was writing in the days before photography and his description of the concretions is an exemplar of the craft of creating a vivid word picture that combines accuracy with memorable metaphor. He made no attempt to account for them though and in fact their origins continued to be a mystery until 1989 when a PhD student showed that they had actually grown inside the sandstone rock taking around 5-10 million years to do so [3].

Miller also noticed that hollows at the base of these low cliffs were filled with what he described as '*a fine quartzose sand formed almost entirely of disintegrated particles of the soft sandstone*.' At the north end of Cleadale is Traigh na Bigil, a singing sand beach, considered by connoisseurs to be the finest of its kind in Britain. To be truthful, this isn't saying much. The sounds can't compare with those produced by the desert sands of Namibia and Oman and some descriptions in the literature of the home grown variety are less than complimentary - 'balefully sibilant' and an 'eerie whining'.

As is often the case though, Gaelic speakers captured the essence in their name for this beach. *Bigil* translates as the 'squeaking of rats or mice' [4]. The sound is produced by scuffing the surface of the sand by walking vigorously without lifting your feet more than an inch or two. The best place to get a good squeak is between the spring and neap high water mark, *and* after the sun has dried out the sand *and* after a shower of rain has cleaned out the smaller particles.

So, if you're in the mood to ponder, three questions might come to mind. Firstly, 'What properties does this sand have that non-squeaky sand doesn't?' Secondly, 'What minerals make up this unusual sand?' and thirdly, 'How is the sound produced?'. The properties were looked in detail by two intrepid pharmacologists in their spare time [5]. They measured the size distribution and degree of roundness of around 500 sand grains from each of 33 singing sand beaches (some hobby!). They also measured the angle of repose, that is, the maximum slope angle of a small pile of sand, which in layman's terms, and this would have been of great interest to my four year old self, means that squeaking sands make good sand castles!

Squeaky sand, Traigh na Bigil, Eigg.

What they found was that a good squeaky sand was made up of grains mostly if not entirely of quartz, all similar in size and if not spherical, at least had their corners rounded off. Now, theories as to how the sound is produced have included friction between salt-coated grains of sand where the salt acts in an analogous way to rosin between a violin's bow and strings, or thin layers of gas released with a percussive effect when the sand is compressed. The front runner though, is that the squeak is caused by friction between clean and more or less rounded grains.

Let's look at this from the other way round first. If the grains are uneven in size and shape then when scuffed they will tend to lock together and any noise is more likely to be a barely audible low-register swiftly cut-off grunt than a squeak. If the grains are more or less rounded though, they can ride over each other unimpeded. In technical language, the grains are working collaboratively to shear along thin slip planes rather than haphazardly through the sand mass.

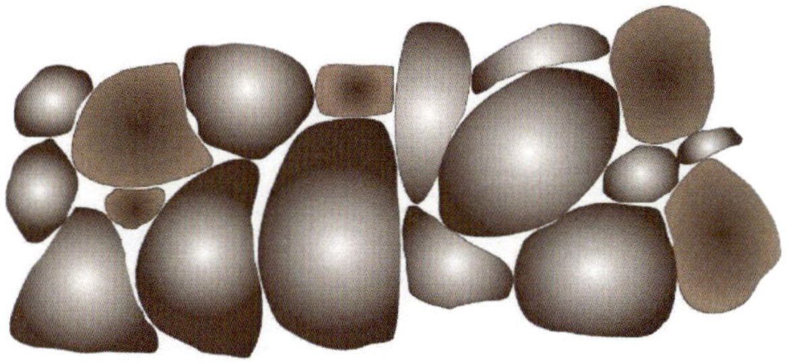

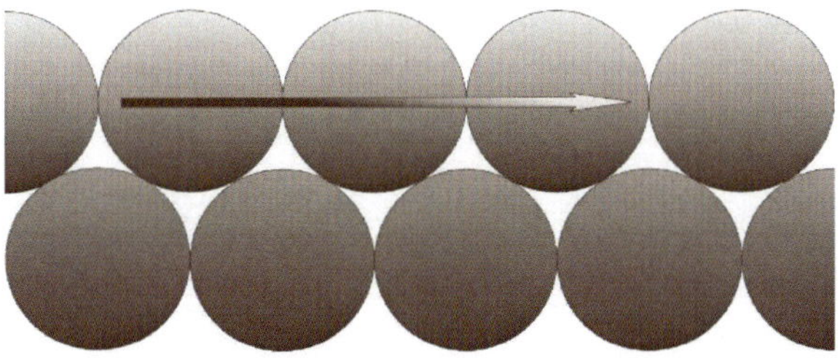

Oddly enough, the large beach next to Traigh na Bigil apparently doesn't squeak, probably because it has a high proportion of shell fragments amongst the quartz. And that makes a neat segue to the next story.

St Andrews, West Sands. This is our favourite dog-walking venue and while Maisie runs around catching tennis balls, or complaining about other dogs using **her** beach while simultaneously sucking up to their owners, I look at the sand, which is pretty much the same as Tentsmuir's - quartz based with a variety of other minerals and ground down rocks. The shallowness of St Andrews Bay however, means that not only is there abundant invertebrate life, but also that it is highly vulnerable in storm conditions. Large wrecks of razor clams, polychaete worms and even deep burrowing otter shells *Lutria lutria* are not uncommon in the winter.

Average waves sort out the finer material.

Sea potato, Echinocardium cordatum

Under the microscope white strandlines are transformed into fabulous collages of minerals and the remains of sea creatures. The clear flakes are muscovite and the dark brown / greeny brown ones are biotite. Both forms are metamorphosed sandy shales, the differences due in part to how close they were to the heat source, e.g. molten granite. Muscovite flakes are often found in sandstones and siltstones while biotite is widely distributed in granite, diorite, gneisses and, perhaps the most likely main source here, metamorphic schists from beyond the Highland Boundary Fault. Mica is a soft mineral but is resistant to water erosion because it behaves like a hydrofoil.

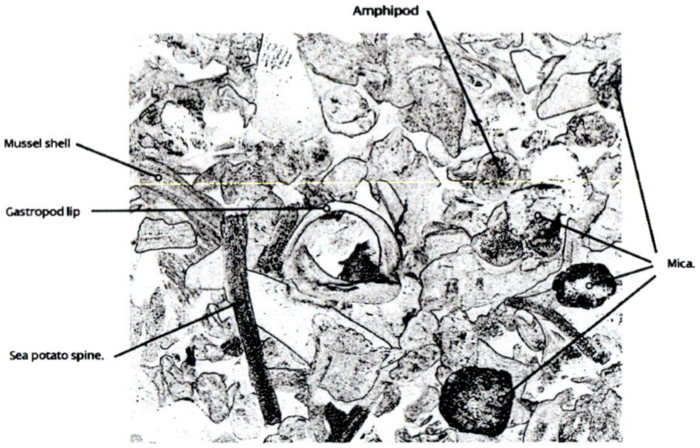

Typical strandline debris, West Sands, St Andrews, Fife.

Amphipods are tiny crustaceans that superficially resemble shrimps but lack the carapace [6]. The living animals aren't curled up like this so I assume that the tissues connecting the segments shrink when the dead animal dries out. Washed up sea potatoes are almost always bald but the living animal is covered with a mat of spines. They spend their lives buried in the sand and use the spines to trap air which helps them avoid asphyxiation.

Conclusion. Well, that just about wraps it up, but I'll finish by mentioning a couple of future sandy projects. First of all, sand is fascinating but challenging. Identifying rocks in the hand is tricky enough but when they're just specks under a microscope it's almost as bad as looking down on a foreign country from an airplane and trying to work out what language they're speaking! Serendipitously though, Rosalind is now doing a course in mineralogy and that's a great help as I work through the sands and gravels collected last summer from several of the rivers and lochs in and around the Tay catchment.

References.

1. *British Geological Survey: British Geological Viewer.* http://mapapps.bgs.ac.uk/geologyofbritain/home.html (Accessed 21/03/18)
2. Hugh Miller (1857) *The Cruise of the Betsey, or A Summer Ramble Among the Fossiliferous Deposits of the Hebrides.* http://www.gutenberg.org/files/28273/28273-h/28273-h.htm#Page_66 (Accessed 21/03/18)
3. Mark Wilkinson (1989) *Sand-hosted concretionary cements of the Hebrides, Scotland.* Thesis submitted for the Degree of Doctor of Philosophy, University of Leicester.
4. The Highland Society of Scotland (1828) *Dictionary Scoto-Celticum: A Dictionary of the Scottish Language, Vol 1*, William Blackwood, Edinburgh.

5. K. Ridgeway and J.B. Scotton (1973) 'Whistling Sand Beaches in the British Isles,' *Sedimentology* 20, pp. 263-279.
6. Peter Hayward, Tony Nelson-Smith and Chris Shields (1996) *Collins Pocket Guide: Sea Shore of Britain and Europe*. Harper Collins.

About the Speaker

Michael Hughes is an unfrocked vivisectionist who discovered that looking after wildlife offered a more satisfying career. After an apprenticeship in the Scottish mountains he and his wife moved down and south to the Gower Peninsula in South Wales where he worked on sand dune National Nature Reserves for thirty years. They both took early retirement in 2005 and moved back to Jennie's home county of Fife, where amongst other things, he polished up his frankly rubbish degree in Zoology by adding a post graduate diploma in heritage interpretation while working freelance. Having finally learnt how to learn properly, geology and geography have become his main interests and he is currently researching and writing a book with the provisional title '*Sand Stories... are more interesting than you might think*'.

What Makes a Great Beach?

Prof. Stephen P. ("Dr. Beach") Leatherman
Laboratory for Coastal Research
Florida International University

The **International Sand Collectors Society (ISCS)** is pleased to present *Prof. Stephen P. Leatherman* as its 2018 SandFest banquet speaker. Internationally-known for his research, and especially for his reputation as "Dr. Beach", Prof. Leatherman has made numerous contributions to earth science, popular literature, and the media.

After completing his undergraduate degree in Geosciences from North Carolina State University, he earned a Ph.D. in Environmental (Coastal) Sciences from the University of Virginia. He has held a variety of positions in academia including Assistance Professor in Geology (Brown University); Director of the National Park Research Unit (University of Massachusetts – Amherst); Professor and Director of the Laboratory for Coastal Research (University of Maryland); and currently, Professor and Director of the Laboratory for Coastal Research (Florida International University). During his productive career, he has authored or edited sixteen books and over 200 journal articles in leading scientific publications. His work has earned him appearances on television, in other media, and even as an expert testifier before the U.S. House of Representatives. The recipient of a host of awards and frequently quoted, Prof. Leatherman will describe his beach rating method* that as have earned him the title "Dr. Beach".

* The following content is based on Leatherman's article, "Rating Beaches," in the *Encyclopedia of Coastal Science,* Springer, Dordrecht, Netherlands (2005), p. 785.

Beaches: the number one recreational destination. Beaches have become the number one recreational destination for Americans and Europeans, and a beach culture has developed worldwide. As a result, people are flocking to the shore in ever-increasing numbers, searching for places where they can escape from urban confinement and everyday pressures. However, with so many choices available, what characteristics do the recreation seekers most value in a beach? Challenged by this question, Prof. Leatherman designed a beach rating survey of fifty criteria and it has been used annually to rate beaches in the United States.

Beach Rating Criteria. Beach rating criteria include physical factors, biological factors, human use, and impacts. Each criterion contains five categories to enable a quantitative rating. The fifty criteria are listed in Table 1, below. Detailed information on how the categories are defined and used for grading may be found in the publication cited earlier.

Table 1. Fifty Criteria Used to Rate the Best U.S. Beaches

Criterion	Description	Criterion	Description
1	Beach width at low tide	26	Pests
2	Beach material	27	Sewage/run-off lines
3	Beach condition or variation	28	Seaweed and jellyfish
4	Sand softness	29	Trash / litter
5	Water temperature	30	Oil / tar balls
6	Air temperature	31	Glass and rubble
7	Number of sunny days	32	Views and vistas – local
8	Amount of rain	33	Views and vistas – far
9	Wind speeds	34	Buildings and urbanism
10	Size of breaking waves	35	Access
11	Number of waves / width of breaker zone	36	Misfits (power stations, dumping)
12	Beach slope	37	Vegetation – nearby
13	Longshore current	38	Well-kept grounds
14	Rip currents	39	Amenities
15	Color of sand	40	Lifeguards
16	Tidal range	41	Safety record
17	Beach shape	42	Domestic animals
18	Bathing area bottom conditions	43	Noise (cars, highways, trains)
19	Turbidity	44	Noise (crowds, radios)
20	Water color	45	Seawalls, riprap, concrete/rubble
21	Floating / suspended human material	46	Intensity of use (overcrowded)
22	Algae in water amount	47	Off-road vehicles
23	Red tide	48	Floatables in water
24	Smell	49	Public safety (crime)
25	Wildlife	50	Competition for free use of the beach

Based on the criteria in Table 1, the title "Number One Best U.S. Beach in 2017" was awarded to Siesta Key Beach on Florida's Gulf Coast. The title at the end of the year is retired to allow others to compete for the next annual title. As a final note, there are a number of awards for best beaches developed by other countries; however, they tend to use fewer quantitative criteria, and few are regarded with the same degree of trust as the more rigorous rating described in this presentation.

More information may be found by visiting the website: http://www.drbeach.org

Workshops

Sand 101: Thoughts for Educators on Teaching With and About Sand

Olga Jarrett
Professor Emerita
Georgia State University

In every outthrust headland, in every curving beach,
In every grain of sand there is the story of the earth.
Rachel Carson

Sand is among the most available and versatile of teaching materials. However, the typical school no longer has a sand box on its playground and sand is not likely to be found in the science supply room. The purpose of this paper is to explore the nature of sand and the many concepts that can be taught through sand.

The nature of sand

Sand is defined as mineral particles of a particular grain size, now standardized at 0.064 mm to 2 mm (.0025 to .079 inches). Generally, scientists work with the metric system but for those more used to thinking in fractions of inches, the ranges are from 1/320" to 5/64", not exactly something that would be easy to measure in a classroom. Digital microscopes are available that operate with a computer and are capable of measuring the size of individual sand grains. For example, I use a Dino-Lite digital microscope. See **Figure 1** below.

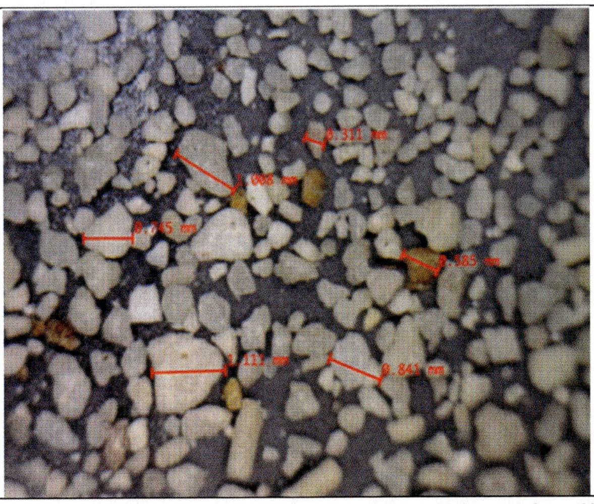

Figure 1. Image with measurements taken with a Dino-Lite digital microscope

Generally, sand can be thought of as grains that are light enough to be picked up by the wind but heavy enough not to remain suspended in the air (Mangimeli, 2007). According to the World Atlas of sands (2011), "in the USA sand is usually classified by grain size into five main categories: very fine (1/16-1/8 mm), fine (1/8-1/4 mm, medium (1/4-1/2 mm), coarse (1/2-1mm and very coarse (1-2 mm)." The coarseness makes a big difference when walking barefoot on a beach. Fine sand feels silky while coarse sand feels as if one's feet are being sandpapered. To sort sand sizes scientifically use sand sieves, a series of nested sieves, each with smaller holes. The materials that get caught by the largest mesh are pebbles, while those that go through the smallest mesh are clay or silt. Haseeb Jamal (2017) describes the process. Sand sieves are a nice addition to the classroom. However, since they are relatively expensive (interlocking sieves sell for over $150), some attempt to collect data on size can be done with kitchen sieves. I do not recommend beach and sand table play sieves.

The composition of sand

In the choice of "animal, vegetable, mineral," sand is definitely a mineral. This excludes other sand-sized particles from being considered sand. For example, granulated sugar and sand-sized spices are not sand. But what about table salt? It is a mineral and is the right size. Could salt be considered a type of sand? That is a mystery for you and your students to argue over. Sand particles can be made up of various minerals. Most beach sand and dune sand on America's coasts are made up primarily of the mineral quartz and is referred to as silica sand. It can also include other minerals such as garnet, zircon, and magnetite. Those with significant inclusion of these other minerals are referred to as heavy mineral sands, since the additional minerals are heavier than quartz. The magnification in **Figure 2** has rose quartz and garnets in it. The minerals that make up sand are those that do not dissolve easily. The beach comber can also find bits of shells and sand dollars but these fragments make up only a tiny proportion of the sand.

Figure 2. Sand with rose quartz and garnets

However, on islands, especially those on or near coral reefs, another type of sand predominates. Referred to as biogenic sand, the beaches are covered with tiny bits of the remains of sea creatures, mineral remains such as sea urchin spines, corals, sea shells, sponge spicules, and foraminifera (skeletons of tiny one-celled creatures). Illustrations of the biogenic sands of Hawaii can be found as part of the Exploring Our Fluid Earth unit from the University of Hawaii (2015). The composition of these sands is very different from the typical mainland beach sand. Instead of quartz, the primary minerals are calcium carbonates ($CaC0_3$). Carbonates are more soluble than silicates that make up the beach sands. The sand in **Figure 3** is from one of the Belizean Cayes.

Figure 3. Sand from a Belizean caye

Having said that all sand is composed of minerals, what I say next may be confusing. Generally sand made from minerals that come from rocks are referred to as mineral sands while sand from minerals made by creatures (sea urchin spines, corals, sponge spicules, or shells) is referred to a biogenic sand. Hence, I will use that distinction in this article.

The origin of sands

Mineral sands are created through weathering. The sands on the Atlantic beaches of the United States came primarily from the Appalachian Mountain range, mountains once taller than the Rocky Mountains (Welland, 2009). As these quartz-rich igneous rocks (primarily granite) and metamorphic rocks (primarily gneiss) weathered, the rocks disintegrated into their separate mineral particles, various combinations of quartz, feldspars, and mica and a few other minerals. The weathering was both physical (as in wind, ice, and rain) and chemical, as plant and animal chemicals as well as acid rain, caused the rock particles to dissolve and break apart. In the mountains, little pockets in the rocks hold small piles of sand-sized particles. These sands are *in situ*, located in the spots where they were made. And these piles hold quite a mixture of sands with various chemical compositions. The feldspars in the granites tend to disintegrate further and eventually become clay. The less hard minerals (including the feldspars) and minerals that are more soluble do not tend to end up as beach sand. Quartz, on the other hand, is hard and long lasting. Mineral sands can also come from volcanoes. Hawaii, Guatemala, and Iceland have beaches covered with black sand from tiny bits of ash and lava. Sometimes these sands are produced very quickly. On the big island of Hawaii, hot lava sometimes flows directly into the ocean. When it does, the lava is shattered into tiny sand-sized pieces, bits that soon end up on the beaches.

Biogenic sands are found near coral reefs, generally on island beaches. These sands have not moved far from where they were made from dead corals, shells, and other pieces of sea creatures. However, sands that are made up of fossils are also referred to as biogenic.

Sand transportation

Water erosion. Sand is moved by erosion. That happens in two ways, by water and by wind. Rain in the mountains removes the piles of sand from where the rock has decomposed and washes them into creeks and then rivers. Both sand and larger rocks such as pebbles, cobbles and boulders (other size categories) move downstream. But sand, being so much lighter, tends to be stirred up by flowing water and move with the current. It can take millions of years, but sand eroded from the mountains can end up on a beach hundreds of miles away.

Wind erosion. Sand also moves by wind. Wind deposited sand is called an "aeolian" deposit. I was once caught in a sand storm in Greece, a storm that made a strong impression on me. Sand can be blown by the wind with stinging force. Sand dunes are made up of wind-blown sand typically located just inland from sandy beaches. The sand piles up because sea to land breezes (known as sea breezes) across the seasons move more sand inland than land to sea breezes carry it back out to sea. And once plants get established on a dune, they help to hold the dune in place. Many dunes are protected and people are forbidden to walk on the dune or disturb the plants.

Figure 4. Dunes on Jekyll Island, Georgia

But sand dunes are also located inland. South Georgia has several large sand dunes. The Google satellite map (**Figure 5**) shows a dune area in Albany, GA (Fossil Sand Dunes… (n.d.)), These dunes are thought to be over 16,000 years old, formed during the last Ice Age when winds from the west blew desert sands into the area. Under water after melting glaciers raised the sea level, these dunes are now about 160 miles from the coast. Dunes are also found along the Flint River and at Little Ocmulgee State Park. Some of these dunes are now being mined commercially.

Figure 5. Dune area in Albany, GA (Fossil Sand Dunes)

The photos in **Figure 6** show layering in a dune on Jekyll Island, Georgia. The wind carries beach sand onto the dune. In this case, the beach sand has layers that are rich in dark heavy minerals such as magnetite and ilmenite blown from the beach, making layers in the dune. [A point of clarification. Having said that dunes are windblown, mounds of sand under water that are moved around by the water are sometimes also referred to as dunes.]

Figure 6. Example of layering in a dune on Jekyll Island, Georgia

Some inland dunes are trapped between mountain ranges. At Great Sand Dunes National Monument in Colorado, melting snow in the Spring brings sand down from the mountains in creeks that dry up during the summer. The winds blow the sand, which is trapped from going anywhere else by two mountain ranges. The sand looks very --- sand colored as shown by a photo in **Figure 7**, left of Great Sand Dunes. But examined under a microscope, the sand looks very diverse as can be seen in the photo in **Figure 8** below. The darkest particles of sand are attracted to a magnet.

Figure 7. Colorado Dunes

Figure 8. Sand from Colorado Dunes

The geometry of sand dunes

News of landslides, mudslides, and collapsing dunes raises the question of their stability. Sand in dunes is mounded up but can get only so steep and then the sand runs down the sides. The angle at which fine grained substances can be piled up is called the angle of repose. Look at the piles of material at construction sites. The angle of repose for sand depends on whether sand is dry or moist and whether sand is fine or coarse. The angle of repose of fine dry sand is 34-35°, of coarse sand is 40°, and of angular gravel is 45°. Damp sand can be made to stand straight up and down for a while. Just picture the walls of a sand castle. But if the castle dries out, the walls will collapse until the sides are at a 34-35° angle. At that point the internal friction of the sand grains equals the force of gravity. But very wet sand has a very low angle of repose, only 15-30°. Picture a sand castle collapsing as the tide comes in.

Because of wind patterns, sand dunes can have various shapes, crescentic, linear, star, dome, and parabolic (Dune, n.d.). Crescentic dunes are crescent or C-shaped and the essentially constant windward side is the open side of the crescent. These dunes are the fastest moving of the dune types. Linear dunes form in lines where the wind changes to blow fairly equally on directly opposite sides of the dune. Star dunes radiate along at least three sides, and they form where the wind changes to blow from several different directions. Parabolic dunes are rather similar to crescentic dunes; but the winds blow outward from the center of the C, moving the tips of the C the fastest. Dome dunes are circular with the wind shifting to come from any side. This is the rarest type of dunes.

Sand Sorting

I mentioned earlier that sand is classified by size of particle. There are several other ways sand is classified: by degree of rounding (are the grains sharp or rounded), by sphericity (how close to little balls the grains are), and sorting (are the grains well sorted --- about the same size or poorly sorted ---a range of sizes?). These classifications can tell a lot about how old the sand is and where it came from. Rounded grains have been rubbed against each other by wind or water. Waves tend to round sand, so sand from beaches with much wave action or long travel in streams and rivers is likely to be more rounded.

Chemistry of sand

Quartz is silicon dioxide with the formula, SiO_2. Silicon makes up 28% of the earth's crust. Only oxygen is more plentiful by weight (47%). Quartz is one of the harder minerals, number 7 out of 10 on the Mohs hardness scale. Topaz, corundum, and diamond are harder but most of earth's minerals are softer. Over the millenia, quartz can get broken into smaller bits or quartz grains can get cemented to make sandstone, but quartz dissolves extremely slowly under normal conditions. It lasts much longer than biogenic sand.

Figure 9. Quartz sand

Biogenic sand is made up primarily of calcium carbonate ($CaCO_3$.), the same mineral that is found in our bones. Egg shells are also formed of calcium carbonate. This is the mineral of limestone, a rock made from sea creature sediments. Limestone gradually dissolves in rain water, which is naturally slightly acidic because it reacts with CO_2 in the air to form carbonic acid, creating caves. Calcium carbonate dissolves in vinegar. The photo in **Figure 10** shows biogenic sand. **Figure 11** is biogenic sand after it has reacted with vinegar to form calcium acetate, water and carbon dioxide.

Figure 10. Biogenic sand

Figure 11. Biogenic sand after reacting with vinegar to form calcium acetate, water, and carbon dioxide.

A fascinating kind of sand found along the coast of Florida is coquina sand. Coquina is a type of rock made of shells cemented together, and large chunks of the rock are found along some beaches (**Figure 12**). As the rocks break up, they mix with the mineral sands on the beaches, creating patches of brownish-orangish coarse biogenic sand among the whiter mineral sands (**Figure 13**).

Figure 12. Coquina rock

Figure 13. Coquina and mineral sand

Categorization of sand

Sand from every beach, creek, and dune is different. Specialists can determine what beach a particular sand came from by analyzing its minerals and knowing which minerals were found in the mountains drained by the creeks and rivers that flow into the sea near that beach. Sand collectors are fascinated by the differences in color and texture of sands found in difference places.

Acquiring sand

This section shows several ways to acquire sand. The easiest is to buy a large bag of play sand from a building supply company. If you are lucky, you will find a broken bag that you can bargain for. You might even convince the company that your classroom is a good cause and that you should get your broken bag free.

Figure 14. Example of a sand collection

Of course, you can use the need for sand as a good excuse to head to the beach. My need for sand has taken me to some exotic places. You can also spread the word to parents of your students and to your friends that you would like sand if they are traveling. Also, I recommend doing a bit of research on line about schools around the world that might be close to beaches or other sand sources. See whether they are interested in becoming pen pals with your class. With or without sand, having pen pals around the world can be a great experience for your students. You can offer to send them local rocks or sand and hope they are willing to send some to you.

Lessons from sand

Teachers can use sand in many ways in the classroom, both to teach about sand and to teach other concepts. Having a stereo microscope would be extremely useful. I recommend rechargeable cordless stereo microscopes so that they can be charged and then taken outdoors. For looking at sand, 20 or 40 times magnification provides sufficient detail while allowing examination of a number of grains at once. There are several good microscopes available for such use including the Cynmar 2X/4X and the Celestron Infiniview that, as of this writing, are reasonably affordable. Once charged up they can be moved anywhere including outside. I generally set up learning stations so only 4-5 students have to share the microscope at one time. If there is no budget for a microscope, try to have at least 10X good quality magnifiers. Here are some ideas for teaching about sand:

1. Have children try to make sand by rubbing rocks together or putting pebbles in a coffee can with a lid and shaking it (expect noise). This is an example of physical weathering. Let them examine the grit they have created with a magnifier or microscope.
2. Collect sand from various places (from family travels or from pen pals) and find photos of those places. Find these places on a map or globe. Create an exhibit.
3. Make sand viewers of the sands you collect. I fold business cards in half lengthwise, cut out a small wedge lengthwise, fasten a strip of Scotch tape on the back across the wedge, spoon a bit of sand on the tape and press down. Pour excess back into the container of sand. (Do not put tape on top of the sand as it will distort view of the sand.) Label the sand with whatever information you have, especially location. But if you know the date and who collected the sand, include that too.

Figure 15. Sand viewing cards

4. Look at the magnified sand to categorize how rounded, spherical, and well sorted it is.
5. If you have black sand, see whether it is attracted to a magnet. Use a strong magnet in a sturdy zip lock bag and drag it through the sand. (The bag keeps the particles from sticking almost permanently to the magnet.) Then pull the magnet out of the bag, being careful not to pick up any of the sand. Allow the sand that was attracted to the magnet to collect on a piece of paper. Later you can make a viewer from it (see #3 above). These sand particles are probably magnetite or ilmenite.

6. Submerge small samples of sand in vinegar (I do this in milk bottle caps). Start with about ½ teaspoon of sand. Add drips of vinegar with a dropper until the sand is submerged. There might be a few bubbles of trapped air in any kind of sand. But if you see bubbles continue to rise, you have biogenic sand. Leave it for a few days to see whether you get calcium acetate deposits (see photo in the chemistry section). If the vinegar does nothing, you have only mineral sand.
7. If you know the composition of your sand and can find some of the minerals that make it up, you can make a matching game. For example, you can have chunks of quartz, garnet, lava, coral, and shells to match with sand samples.
8. Sieve the sand and collect what does and does not go through the sieve (or a collection of sieves). Measure or weigh the various size particles. Make a graph of the weights and volumes of various sizes as fractions of the total.
9. Show how sand ripples are created. http://sand.xboltz.net/ripple_marks.pdf. Sand ripples are like tiny dunes under water.
10. Show the limits of the heights of dunes. Measure the angle of repose of various sands and other granulated materials. Pour clean sand through a funnel, making a pile on a piece of paper. You should be able to read the angle at around 34 or 35 degrees. Try other materials as listed on this website: https://en.wikipedia.org/wiki/Angle_of_repose and various methods for measuring angle of repose. I used a protractor with its center lined up with the edge of the pile of sand. You will need to line up the protractor on the edge of the table and get down at eye level to accurately see the angle of the edge of the pile of sand.

Using sand to teach other concepts:

1. Children are often taught that a solid keeps its shape but a liquid takes the shape of its container. So, is sand a liquid or a solid? It is a solid. But it behaves like a liquid.
2. Use sand for weighing and measuring. How do a cup of sand and a cup of pebbles compare? Do all sands weigh the same for the same volume?
3. Spread a thin layer of sand in a tray and have children draw letters or pictures in it.
4. Various ethnic groups make pictures from sand, making sand art appropriate for teaching history and geography. **Figure 16** shows Tibetan monks making a sand mandala. **Figure 17** is a Navaho sand painting. Traditionally, these sand pictures are not kept but are swept into water or into the wind when finished. They serve religious purposes. **Figure 18** is modern art with natural sands from Providence Canyon State Park in Georgia that I glued onto cardboard.

Figure 16. Tibetan monks making a sand mandala

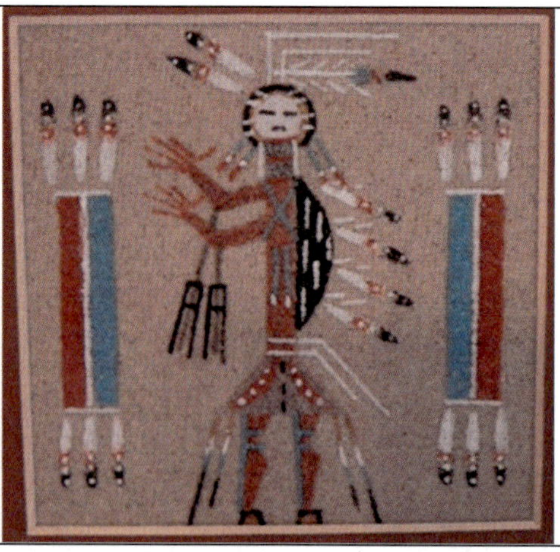

Figure 17. Navajo sand painting

Figure 18. Modern sand art

References

Carlisle, M. W. (1993). *Let's investigate soft, shimmering sand: Historical and scientific facts about nature and human inventions*. Hauppauge, NY: Barron's. [Informative children's book.]

Clary, R. & Wandersee, J. (2011, Summer). A scientific world in a grain of sand: Investigating local sand samples on a shoestring budget. *The Science Teacher, 78*(5), 29-33. [This article has excellent ideas for high school teachers on using sand in teaching earth science.]

Dune (n.d.) National Geographic. Retrieved from https://www.nationalgeographic.org/encyclopedia/dune/

Edwards, L., Ambrose, J., & Kirkman, L. K. (2013). *The natural communities of Georgia.* Athens, GA: University of Georgia Press. [This book discusses natural communities and the soil environments that support them.]

Exploring Our Fluid Earth unit from the University of Hawaii (2015). Retrieved from https://manoa.hawaii.edu/sealearning/sites/default/files/M1U5-Table5.8.%20Common%20components%20of%20biogenic%20sand.pdf

Fossil sand dunes in Albany, Georgia. Retrieved from http://matthewinatl.com/pasado/fossil_dunes.php

Gallant, R. A. (1997). Sand on the move: The story of dunes/ New York: Franklin Watts. [Informative children's book.]

Greenberg, G. (2008). *A grain of sand: Nature's secret wonder*. Minneapolis, MN: Voyageur Press. [Beautiful book of sands under the microscope.]

Greenberg, G., Kiely, C. & Clover, K. (2015). *The secrets of sand: A journey into the amazing microscopic world of sand.* Minneapolis, MN: Voyageur Press. [Beautiful book of sands under the microscope.]

Jamal, H. (2017). Sieve analysis & particle size analysis. About Civil Engineering. Retrieved from https://www.aboutcivil.org/Sieve-analysis-and-soil-classification.html

Mangimeli, J. (2007) Geology of sand dunes. Retrieved from https://cms.imr.nps.gov/whsa/naturescience/upload/Sand%20Dune%20Geology.html

Neal, W. J., Pilkey, O. H., & Kelley, J. T. (2007). Atlantic coast beaches: A guide to ripples, dunes, and other natural features of the seashore. Missoula, Montana: Mountain Press Publishing Company. [Informative book on what can be found on Atlantic beaches.]

Pilkey, C. O. & Pilkey, O. H. (2016). *Lessons from the sand: Family-friendly science activities you can do on a Carolina beach.* Chapel Hill: University of North Carolina Press. [Many science experiments that can be done at beaches all over the world.]

Prager, E. J. (2000). *Sand.* Washington, DC: National Geographic Society. [Very informative and readable book by a geologist on the origin and nature of sand.]

Steinmetz, G. (2012, November). Sailing the dunes. *National Geographic*, 94-109 http://waynesword.palomar.edu/ww0704.htm#Introduction (about dunes)

World Atlas of sands (2011). Retrieved from http://www.sand-atlas.com/en/sand/

Welland, M. (2009). *Sand: The never-ending story.* Berkeley, CA: University of California Press. [Excellent children's book on various types of sand, what sand looks like close up and how sand is created.]

About the Workshop Presenter

Olga Jarrett is Professor Emerita of science education in the Department of Early Childhood and Elementary Education at Georgia State University. She has long been a sand collector and member of the International Sand Collectors Society and has used sand in teaching about earth sciences in her science education courses. She has also given workshops on teaching with sand for Georgia Science teachers Association and National Science Teachers Association conferences. Her research includes studies of science majors, educators, and geologists on childhood experiences that develop interest in science. She is currently co-editor of the Georgia Mineral Society (GMS) newsletter and a member of the GMS Sand Section.

Sand Pen Physics: A Nutty Puzzle

Lori Carter
GMS and ISCS Member
Roswell, GA

Introduction. When sand and science and art come together it is a great day for all! The purpose of the sand pen workshop is to make a pen that can be used as an actual writing pen while also presenting a physics puzzle.

Making the pen. For this workshop we use large, hollow, plastic pens that are normally meant to be filled with layers of different colors of sand. Instead of colored sand, we fill the pens with white sand from Spruce Pine, NC. The sand is angular, almost pure silica, and fairly uniform in size and shape. A steel ball bearing discarded during auto part manufacturing is also added with the sand, and the lid is screwed on tightly.

The puzzle. Once the pen is ready, the challenge is to get the ball bearing to move from one end of the pen to the other. People try to roll the ball while tilting the pen; they try to turn the pen around and around while holding it horizontally; they try various methods to no avail.

The solution. Eventually one learns that by shaking the pen vertically, the ball rises up the column of sand little by little until it reaches the top. But why? According to physicists who study this sort of thing, the answer lies in a curious physics conundrum called the "Brazil Nut Effect". People who crave Brazil nuts know that the objects of their desire can be coaxed to the top of a can of mixed nuts by simply shaking the can up and down. Physicists are still studying this effect, so we will examine what is known at the time of this writing. There are at least two potential explanations of the Brazil Nut Effect – *percolation* and *convection.*

Percolation: As the pen is moved up, the sand and the ball bearing are thrust upwards. As gravity pulls them back down, the smaller sand particles fill voids below the ball bearing. As the process is repeated, the ball bearing goes further and further up the tube.

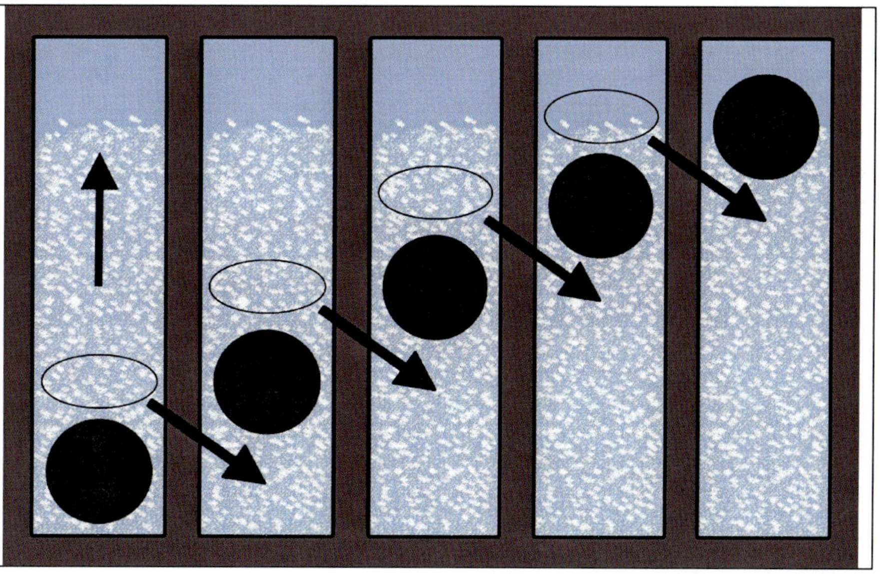

Convection: As the pen is shaken, particles in the center of the tube go up, and particles on the sides go down. The ball bearing moves up via convection, but for some reason does not go back down, possibly because the ball is too big for the convection current.

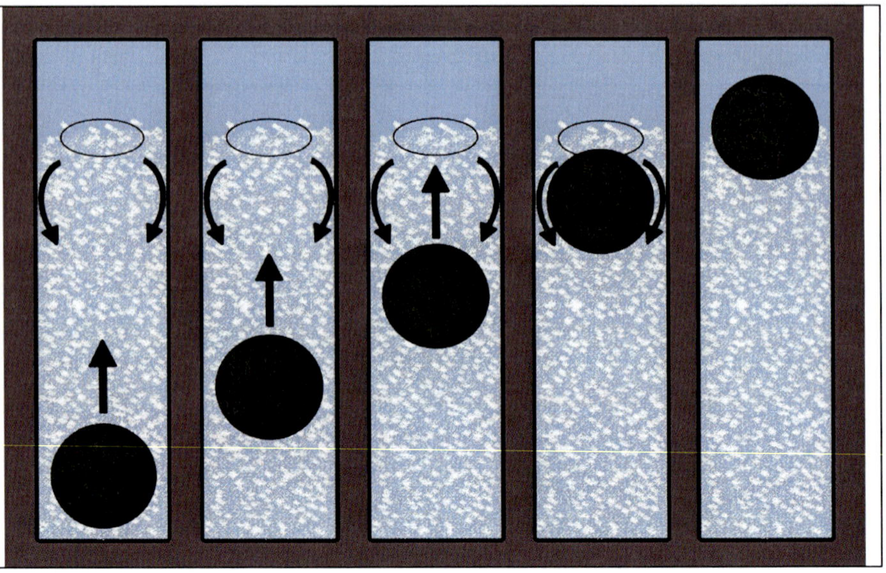

The Brazil Nut Effect has also been observed in places where the ground freezes and thaws. As underground water freezes, the ice pushes particles up. When the ice thaws, the particles fall back down. The smaller particles fill in below the larger particles, so rocks are pushed up, much to the chagrin of farmers who thought they had cleared their pastures of the pesky stones.

Conclusions. This simple sand pen workshop introduces people to a physics problem that is still being studied. The Brazil Nut Effect is more complex as one delves deeper into the physics. Interstitial space, density, friction, gravity, fluid dynamics, granular convection – many things must be considered, even how the effect has different results in vessels of different shapes and in a vacuum. So, when you make a sand pen, not only will you realize you now have a cool pen to write with, but you will also realize you have opened a physics can of worms – actually, mixed nuts.

About the Workshop Presenter

Lori Carter enjoys all sorts of sand topics. Like Alice chasing the white rabbit, she often finds herself down a rabbit hole seeking answers to the mysteries sand can engender. She visits many schools as the "Sand Princess" to teach students about geology by looking at and thinking about sand. During these STEM nights, she readily and happily tells students that one of the best things about science is that we don't know everything. The quest for knowledge is as important as knowledge itself.

Carbonate Grains in Sand
Identifying Biogenic Bits and Non-skeletal Grains

Kate Clover
ISCS Member and Sand Collector and Curator, Life-long Educator,
Retired Museum Professional, Saint Paul, Minnesota

Introduction. In this workshop, we will focus on the common marine invertebrates whose carbonate skeletons we find in sand including mollusks, sea urchins, barnacles, and corals. We'll discuss **morphology**, the structure and form of those organisms and see examples in macro samples. We'll also consider **taphonomy;** this is the study of how organisms break down after death and how chemical, physical and/or biological processes transform the skeletal remains from whole into grains the size of sand.

We will study common biogenic species using the Playa Cason, Dominican Republic photo (see below), other sands, and macro-sized specimens. If we understand the original morphology and how things break down, we can more likely identify the fragments after they have disarticulated or been abraded.

The diversity of grains in this sand reflects a rich marine ecosystem. Included here: Urchin parts, forams, molluscs, bryozoans, sponge spicules, halimeda, coralline alga, worm tubes, and rocks.

Playa Cason, Samana, Dominican Republic Photo by Leo Kenney

Biogenic Species Commonly Found in Sands

Corals. Reef building corals thrive in shallow marine environments and are the most important carbonate producer in tropical seas. Some corals create colonies while others are solitary organisms. Learning coral anatomy and being able to describe the coral calyxes (cups) on their skeletons helps with species identification. Because of the diverse morphology of coral species and because it is often fragmented, identification can be difficult.

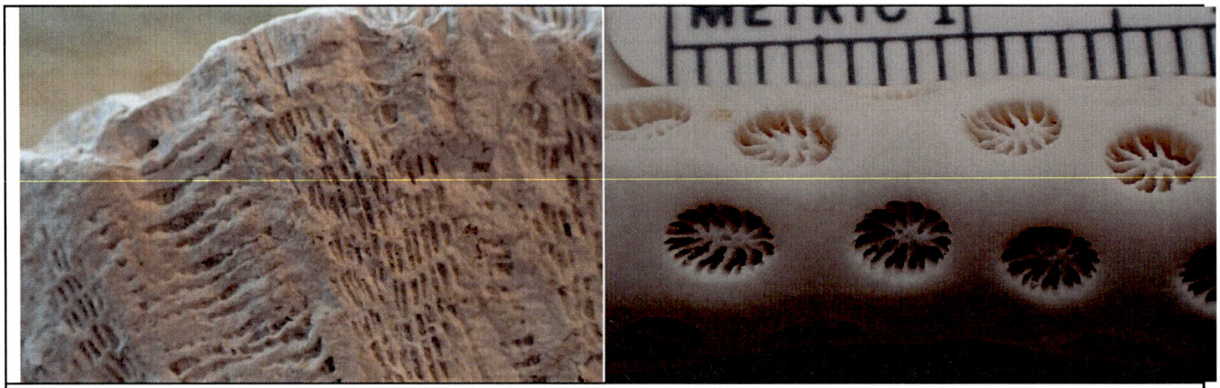

Left: In cross section, coral shows ladder-like growth channels.
Right: This coral sample shows inverted coral calyxes (polyp cups).

Molluscs. Molluscs are one of the largest and most familiar groups of invertebrates. The phylum includes 1) bivalves: clams, scallops, mussels and oysters; and 2) gastropods— the snails. There is no "typical" mollusc; their forms show great diversity. Their shells are comprised of one of two minerals: calcite or aragonite. Calcite specimens are chalkier; the aragonite forms are shinier. Some molluscs have layers of both calcite and aragonite.

Left: Macro-sized gastropod fragments illustrate their coiled internal morphology. The columella (the central rod) is the most durable; the apex (the pointy end) is the first part of the shell to form.

Left center: This tightly-coiled gastropod, a Turritella snail survives abrasion in the oceans and is frequently seen in sands.

Right center: These macro-sized bivalve fragments show angular and rounded edges, plus growth line, grooves and curvature; some are shiny, others chalky looking. The edible Blue Mussel (lower right) is common in temperate to polar intertidal areas.

Right: Bivalves have two valves (shells) that hinge close; the interlocking teeth on the hinges prevent slippage. The hinge section of the bivalve is the first to form and is the most robust part. Noting the details of the hinge and teeth helps in identification.

Echinoderms. The echinoderm family includes sea stars, sea urchins, sand dollars and crinoids (see fossil section). All echinoderms exhibit five-fold symmetry. Sea Urchins do not have rays (arms) like sea stars, but they have an internal shell called a test that consists of five calcareous plates that fit together. Spines cover the body; they can be used for locomotion, burrowing, protection or food gathering. The cupped end of the spine attaches to the test by muscles and connective tissue.

Left: Sea Urchin test (the body) with spines attached.

Middle: Sample includes three different sea urchin bits. 1) The red rod is an urchin spine. 2) The bumpy green fragment shows the ball joints, the spots where spines attach to the test. Larger spines attach to the larger balls, smaller spines to the smaller ones. 3) The fragment (lower right) with the parallel row of pores shows the interior of an urchin test. **Cap Lardier, Côte d'Azul, France** (Photo by Leo Kenney)

Right: In cross section (and highly magnified), an urchin spine illustrates radial symmetry and resembles a mandala. **Maui, HI** (Photo by Gary Greenberg)

Barnacles. Barnacles are odd crustaceans. In their larval stage, they are free floating, but as adults they permanently attach themselves or burrow into living or non-living substrates. Their bodies consist of a ring of six carbonate plates. Two additional plates, the tergum and scutum, at the top entrance are movable and allow the animal to feed. These plates close up when the animal withdraws. The tergum and scutum seem to be the first to disarticulate from the body after death. All the plates vary in form making identification challenging.

Left: Macro examples of barnacles missing their "mouth plates" –the scutum and tergum.

Center: Grains showing parallel channels are the wall plates of barnacles. Other grains with similar patina are likely barnacle plates too. **Lima, Peru** (Photo by Leo Kenney)

Right: A highly-magnified inner side of a barnacle scutum, one of a plates that covers the barnacle's "mouth" opening. **River Tyne, North Sea, England** (Photo by Carol Keily)

Forams. Forams (short for foraminifera) are single-celled chambered organisms related to amoeba. They secrete calcium carbonate tests or shells and exhibit intricate and beautiful forms. Scientists have identified over 40,000 different modern forams and about the same number in the fossil record. Identification can be challenging.

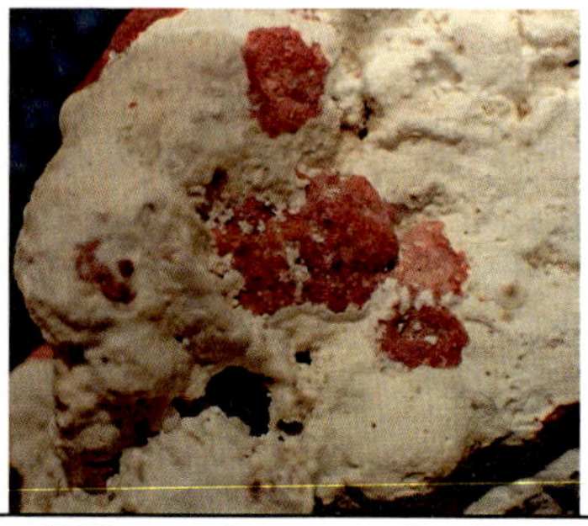

Left: This sample includes at least four different forams, plus bryozoans and urchin spines. **Hobuck Beach, WA (**Photo by Leo Kenney)

Center: The red blobs are the encrusting foram, *Homotrema rubrum* common in the Bahamas. When these fragments loosen from their hard surface, they accumulate in the sand and impart a pink hue to the beach.

Right: Pink sand shows the high concentration of the foram, *Homotrema rubrum.* Mesh-looking grain is a bryozoan. **Bodrie Beach, Corsica, France** (Photo by Leo Kenney)

Bryozoans. Bryozoans (or moss animals) are colonial organisms. They take on different forms: encrusting, branch like, button-like or fenestrate (resembling a window screen). The microscopic animals, called zooids, live inside the oval-shaped compartments. Often confused with corals, the oval zooid opening sets them apart. About 4000 living species are known, and they are found from the poles to the tropics. The group dates back to the Cambrian period (~540 million years ago) and has left an impressive fossil record.

Left: The oval-shaped opening on these rods once housed zooids, the living member of a bryozoans colony. Note the chalky nature of these rods. The honey-combed black fragment is also a bryozoan (and likely a fossil). **Hakampo Beach, Taeanhaean National Park, South Korea** (Photo by Leo Kenney)

Right: Fossil bryozoan and gastropods, Middle Miocene in age (15 million years old). **Savigne-sur-Lathan, France** (Photos by Leo Kenney)

Marine Plants which Form Carbonate Grains. Algae are important sources of biogenic carbonate sediments in marine environments. Two types of algae are carbonate producers: red algae (rhodophyta) and green algae (chlorophyta).

Red Algae (rhodophyta) are calcified plants that are found as crusts on rocks and other surfaces, as nodules or branches of cylindrical or flattened leaf-like segments.
Coralline algae (rhodophyta) are generally smooth and do not show the distinctive calyxes (polyp cups) found on coral or the oval zooid openings seen on bryozoans. Depending on the location, wave-polished grains can appear shiny or weathered fragments can appear chalky. Color can range from white to rusty orange.

Left: Coralline Algae and a few gastropods. Sections of coralline alga are often confused with coral, but coralline surfaces are smooth and show no polyp or zooid openings. Coralline algae are indicators of an offshore reef. **Galway, Ireland** (Photo by Leo Kenney)

Right: Red Coralline alga consists of calcareous, branching, segmented fronds. Segments are jointed. Color varies from purple, red, pink to yellowish with white knuckles and white extremities. This is a typical alga of rock pools on the middle to lower shore. Found from the north Atlantic to Argentina, also around Japan, and Australia.

Green Algae (chlorophyta) have calcified stems and branches (often segmented). The most common green alga is *Halimeda*, a tropical form with erect blades of articulated, disc-shaped sections.

Left: Fresh *Halimeda* from Andros Island, Bahamas. This alga is common in warm shallow water and starts to calcify while alive. As *Halimeda* easily breaks down with intense wave action, it is only found near its source.

Right: The large flake (center-right) is *Halimeda*. Notice the joint where the segment attached. Sand with lots of *Halimeda* is known as "Oatmeal Sand." Also pictured are forams and an urchin spine. **Andros Island, Bahamas**. (Photo by Leo Kenney)

Fossils. Marine organisms are found as fossil in sands. Many species have contemporary counterparts while other species are only found in the fossil record. The occurrence of marine fossils in a sand or rock indicates the location supported a marine environment in ancient times.

Left: Fossil shark teeth from New Mexico are evidence of marine life in the Western Interior Seaway during the late Cretaceous (~100 million years ago). The sea split North America into two landmasses. **Shark Tooth Ridge, NM** (Photo by Leo Kenney)

Right: Fossil crinoid rings and columns from the Ordovician seas (~450 Million years ago). Crinoids are echinoderms related to sea urchins. **Byrdtown, Tennessee** (Photo by Leo Kenney)

Non-Skeletal Contributors to Carbonate Sediment

Ooids. Ooids are spherical bodies of calcium carbonate less than 2 mm in diameter. They have an internal structure of concentric layers that form by the precipitation of calcium carbonate around a nucleus in a super-saturated carbonate environment. Ooids form in shallow marine environments today: the Bahamas, on the Great Barrier Reef, and in Great Salt Lake, Utah. Ooids are also found in the fossil record as oolitic limestone.

Left: Ooids from Joulters Cay, 10 miles off the north shore of Andros Island in the Bahamas. The nucleus of these ooids is fecal material from brine shrimp that live in the shallow water amongst the ooids. Birds are plentiful on this cay and feed on the shrimp.

Middle: Joulters Cay is expansive and is made almost entirely of ooids. The ebb and flow of the tides along with the currents keeps the ooids in near constant motion, ever growing.

Right: Map of **Andros Island, Bahamas**

Conclusions: In seas around the world, the marine ecosystems vary; and for me, it's been a rewarding experience studying the composition of beach sand. I've learned much about marine invertebrates and the myriad of ecosystems they can inhabit. I've learned about the geology of the nearby landmasses, the marine flora, ocean currents and tides, the geography, paleontology and sometimes history of the region. I've found it fascinating.

Learning to identify the biogenic grains in sand takes lots of experience whether looking under the microscope or at photos of magnified sands. Some grains are easy to identify, others—because of the wear and tear they experience after death, can keep you guessing for a long time. As you walk the shore, notice the macro-sized samples along the swash lines; these will give you clues about the fine bits you'll discover in the sand.

References:

Moore, Lalicker and Fischer (2004) *Invertebrate Fossils* McGraw-Hill Publishing
Pechenik (2010) *Biology of Invertebrates* McGraw-Hill Publishing
Kozloff (1990) *Invertebrates* Saunders College Publishing

Web Links:

Corals web links:
https://oceanservice.noaa.gov/education/tutorial_corals/welcome.html
https://odealab.github.io/ccsig/

Shell web links:
http://www.mba.ac.uk/sites/default/files/jase/shell_key_estuary_commonspecies.pdf

Barnacle web link:
https://crustacea.academic.ru/297/carinolateral

Foram web link:
http://www.foraminifera.eu/queryrw.php?family=&genus=&species=&age=&area=North+Sea&imamo=&aktion=suche

---

About the Workshop Presenter

Kate Clover has collected and studied sands for over 30 years. In 2015, she co-authored *Secrets of Sand* with Gary Greenberg and Carol Kiely. She also co-publishes the *Splendid Sands Calendars* with Leo Kenney and Carol Hopper-Brill. Kate holds a BS in Geology and Technical Communications from Michigan Technological University. Her graduate work earned her a teaching certificate from the University of Wisconsin/ River Falls, and she's studied marine science through Clemson University and on Andros Island, Bahamas. At the Science Museum of Minnesota from 1990 – 2016, she served as a geo-science educator and exhibit program-developer. Retired in 2016, Kate continues to study sand, write about geology, travel and volunteer around the Twin Cities. She also has given talks/workshops on sand at the McCrone Research Institute, the Science Museum of Minnesota and the Minnesota Geological Society. She is a long-time ISCS member.

Credits: Photography by Kate Clover unless otherwise noted. Special thanks to Leo Kenney, Carol Hopper Brill, Carol Kiely, and Gary Greenberg for their beautiful photography, creative collaborative work and inspiration over the years.

In situ Workshop: Sand and Sand Phenomena

Bob Jarrett
GGS, GMS, and ISCS Member
Decatur, GA

Abstract
The field trip group will visit at least four major locations on Jekyll Island to: 1) examine and discuss the dynamics of sand occurrence and movement, 2) obtain water- and air-deposited sand samples for collections, 3) look for and observe relationships between biota and sand, 4) note a variety of dune morphologies (shapes), 5) learn about the interior "construction" of a barrier island, and 6) briefly consider how "heavy minerals" become concentrated and find themselves located far inland at higher elevations. The latter material supports the heavy mineral mine trips.

General Map
Here you see the areas to which we will be going; first to the northern end.

General Information

Our mid-afternoon arrival will be at low tide providing excellent access. We will have at least three (maybe four) hours of available time before the tide is too high to be comfortable on the beaches at which time it will also be dusk in late September.

All walking will be relatively easy (no sharp elevation changes) other than boardwalk steps, but loose sand in places might be a bit problematic for a few trip members. Near the water is solid.

A water fountain and rest rooms are available at the first stop, but water and necessary snacks should be carried, at least in the vehicles.

Wear shoes suitable for walking in sand and some wet-ish spots - we should not need to slog through water, unless it is raining. Check the weather for need to carry rain gear. Carry and use sun block. Have mosquito repellant handy. Carry: camera, loupe, magnet, and zip lock bags.

Stops 1-4, North End, approx. 1.5 hrs.

Stop 1 is the parking lot for the fishing pier. Try to park under the trees off the blacktop - cooler.

Stop 2 is the obvious set of pavilions, water tap, and rest rooms.

- We'll congregate as best we can.
- We'll have two short lecturettes to get us stated:
    - Formative forces at work at the north end of Jekyll (Bob & Olga Jarrett); and
    - Native American archaeology of the coastal region (Kate Clover).

Stop 3 is the Clam Creek bridge to briefly view and discuss the tidal wetland which could open at the other end to create another island.

Clam Creek bridge; tide ebbing

N. Jekyll ebb tide

Stop 4 is at the end of a beach walk to look for north end sand samples (especially dark heavy minerals layer in the beach), dig a trench to view sand layering, observe surface patterning, discuss barrier island southerly drift, and look for beach erosion evidence.

Nature's art in dark minerals

Nature ties a knot using water and sand media

Travel about 10 minutes to:
Stop 5a. Turnoff at Beachview Drive N. and Dexter Lane to get us to **Stop 5b.** Park on streets, approx. 0.5 hrs.

- This is the beach access point for viewing some of the first exposures in untold thousands of years of the Pleistocene sediments (approx. 15,000 years old), peat, and stumps.
- These underlie the Holocene (less than 10,000 years old) dunes and beach.
- These Pleistocene deposits are the "core of the island" exposed by Hurricane Matthew in October 2016.
- Samples of the organic material and beach sand can be taken to represent a mid-island (North to South) context.

Rip-rap and exposed Pleistocene deposits	High-high tide erosion over rip-rap

Travel about 15 minutes to:

Stop 6a. Turnoff at South Loop Trail / Beachview Drive at Jekyll Island Soccer Complex and onto an unnamed street to get us to parking and Stop 6b, Glory Beach Boardwalk, approx. 1.0 hrs.

- Parking and entry to the southeastern beach
- Sample sands
- Look for bioturbation in cutaway dunes and biota markings in sand
- Dig a trench to view sand layering
- Observe surface patterning
- Observe and discuss various ripple forms
- Examine aspects of dune forming and dune destruction
- Possibly capture aeolian (wind-blown) sand. The latter can be compared to water-deposited sand for composition, size, and shape.
- Make 3D sand picture of sand dunes – or at least obtain materials to make at home.
- Look for strange phenomena

Wind deposition and sand saltation

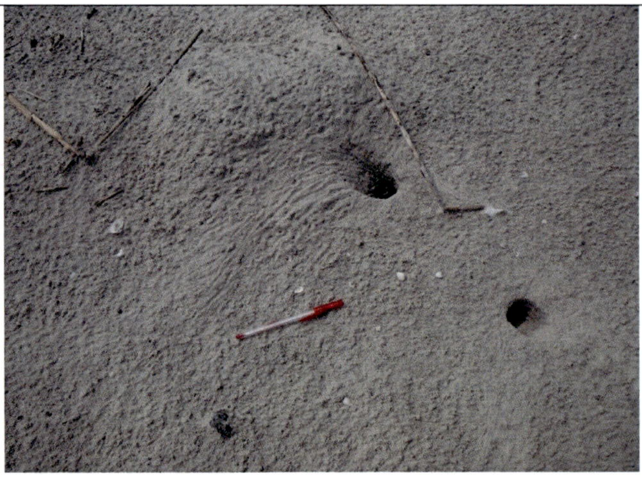

Crab house cleaning and rings in back door

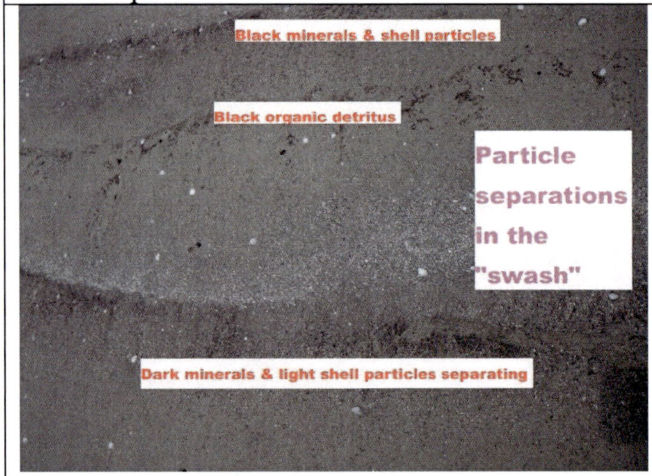

Swash separation of minerals

South spit, mineral layering, less intense

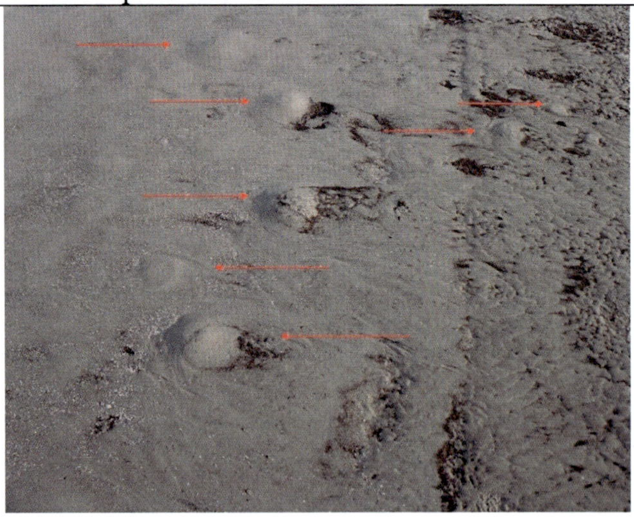

Sand bubbles

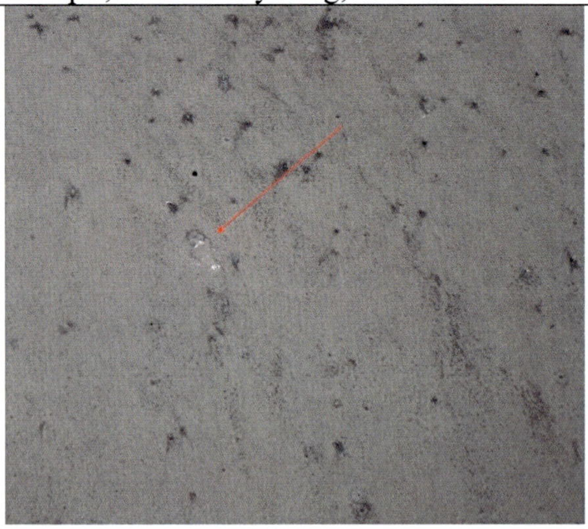

Worm pumps water and worms ejected magnetite

And many, many more interesting phenomena to watch for.

About the Speaker

Bob Jarrett first remembers, at age 4, a large pile of sand in his neighborhood in which kids tried to find tiny "Indian beads," small cylindrical fossils. Eventually, sand, as aquifers for water supplies, became important in his public health/environmental engineer paid and pro bono work in the US and abroad. While working for WHO, he advised water authorities, health ministries, and water engineers in the English-speaking Eastern Caribbean Islands and Belize. He made several trips with non-governmental organizations to train local nationals and advise on drilling safe water wells - in sand. During and after getting a post-retirement MS in geology for the fun of it, he has joined his wife Olga in making many presentations about sand and conducting workshops on sand activities for teachers to use in classes. He is the 2018 president of the Georgia Geological Society, co-editor of the Georgia Mineral Society *Tips-and–Trips* newsletter, Historian for GMS, and a GMS Sand Section member; and he holds memberships in: ISCS, Atlanta Geological Society, Geological Society of America, Clay Minerals Society, and water related organizations.

Notes

Made in United States
North Haven, CT
04 November 2026

21728940R00060